Lecture Notes in Civil Engineering

Volume 243

Lecture Notes in Civil Engineering (LNCE) publishes the latest developments in Civil Engineering - quickly, informally and in top quality. Though original research reported in proceedings and post-proceedings represents the core of LNCE, edited volumes of exceptionally high quality and interest may also be considered for publication. Volumes published in LNCE embrace all aspects and subfields of, as well as new challenges in, Civil Engineering. Topics in the series include:

- Construction and Structural Mechanics
- Building Materials
- Concrete, Steel and Timber Structures
- Geotechnical Engineering
- Earthquake Engineering
- Coastal Engineering
- Ocean and Offshore Engineering; Ships and Floating Structures
- Hydraulics, Hydrology and Water Resources Engineering
- Environmental Engineering and Sustainability
- Structural Health and Monitoring
- Surveying and Geographical Information Systems
- Indoor Environments
- Transportation and Traffic
- Risk Analysis
- Safety and Security

To submit a proposal or request further information, please contact the appropriate Springer Editor:

- Pierpaolo Riva at pierpaolo.riva@springer.com (Europe and Americas);
- Swati Meherishi at swati.meherishi@springer.com (Asia - except China, and Australia, New Zealand);
- Wayne Hu at wayne.hu@springer.com (China).

All books in the series now indexed by Scopus and EI Compendex database!

More information about this series at https://link.springer.com/bookseries/15087

Elham Maghsoudi Nia ·
Iman Farshchi · Lin Yola ·
Mokhtar Awang
Editors

Sustainable Development Approaches

Selected Papers of AUA and ICSGS 2021

Springer

Editors
Elham Maghsoudi Nia
Management in the Built Environment
Delft University of Technology
Delft, The Netherlands

Iman Farshchi
Faculty of Engineering
and Built Environment
MAHSA University
Jenjarom, Selangor, Malaysia

Lin Yola
School of Strategic and Global
Studies Building
Universitas Indonesia
Jakarta, Indonesia

Mokhtar Awang
Department of Mechanical Engineering
Universiti Teknologi Petronas
Seri Iskandar, Perak, Malaysia

ISSN 2366-2557 ISSN 2366-2565 (electronic)
Lecture Notes in Civil Engineering
ISBN 978-3-030-99978-0 ISBN 978-3-030-99979-7 (eBook)
https://doi.org/10.1007/978-3-030-99979-7

This Springer imprint is published by the registered company Springer Nature Switzerland AG
The registered company address is: Gewerbestrasse 11, 6330 Cham, Switzerland

Preface

This book presents a compilation of research works of 2021 AUA Academic Conference in conjunction with the 5th International Conference on Strategic and Global Studies (ICSGS) with the big theme of 'Global Strategies for a Resilient and Sustainable Post Pandemic World Towards a Better Future for All'. The academic conference was organized by School of Strategic and Global Studies, Universitas Indonesia, in collaboration with the Asian Universities Alliance (AUA) on 26–27 October 2021. There is no doubt that the pandemic crisis threatens all aspects of human life. The 2021 AUA and ICSGS Academic Conference invited academics (lecturers, undergraduate and postgraduate students), scientists, practitioners and policymakers from various specializations related to strategic global and sustainable development studies to share their inter/trans/multidisciplinary research findings and professional experiences to impact the post-pandemic world significantly. It provided a platform for an intellectual dialogue on the crucial topics related to post-pandemic crisis risks, impacts and mitigation efforts. More extensive mitigation is needed to achieve alertness to prepare ourselves for possible future pandemics and create a more resilient and sustainable world for everyone. The editor board of the proceeding would like to express the utmost gratitude and appreciation to all expert reviewers in the technical team for making this volume a success.

Contents

Analysing On-Time Performance of Public Bus Service at Northern Peninsular Malaysia

Nor Najwa Irina Mohd Azlan, Shuhairy Norhisham, Muhammad Fadhlullah Abu Bakar, Nur Syafiqah Mohd Shkuri, Wan Ahmad Faiz Wan Mohd Fauzi, Siti Aliyyah Masjuki, Mohd Zakwan Ramli, and Sarah Shaziah Samsudin

Abstract Among public transportation available across the countries, the motion of buses is full of uncertainties because of the traffic congestions, delays in the timetable, and other incidents. But public bus benefits people in many ways including minimising parking problems, and energy consumptions from individual vehicles. The purpose of the study is to analyse the on-time performance of public bus service specifically at capital cities of Kangar, Ipoh, Georgetown, and Alor Setar in Northern Peninsular Malaysia. The method used to analyse the on-time performance of the public bus is by recording the time of early and delay for each route of the bus. Then, on-time performance percentage can be estimated and specified to a quality of service (QOS) provided by Transit Capacity and Quality of Service Manual 2013 (TCQSM 3rd Edition). The result of the study found that QOS at Kangar and Ipoh fall at C which the punctuality of the bus service is debatable. While the QOS at Georgetown and Alor Setar fall at E showing bad perceptions and low reliability of the service to citizens. The role of authorities and stakeholders is very important to ensure the quality of service of these capital cities at the highest level to capture the user's trust.

Keywords Public bus · Northern Peninsular Malaysia · On-time performance · Quality of service · Public transportation

---

N. N. I. M. Azlan (✉) · S. Norhisham · M. F. A. Bakar · W. A. F. W. M. Fauzi · M. Z. Ramli · S. S. Samsudin
Department of Civil Engineering, College of Engineering, Universiti Tenaga Nasional, 43000 Kajang, Selangor, Malaysia
e-mail: najwaazlan94@gmail.com

S. Norhisham · N. S. M. Shkuri · M. Z. Ramli
Institute of Energy Infrastructures (IEI), Universiti Tenaga Nasional, 43000 Kajang, Selangor, Malaysia

N. S. M. Shkuri
College of Energy Economics & Social Sciences, Universiti Tenaga Nasional, 43000 Kajang, Selangor, Malaysia

S. A. Masjuki
Department of Civil Engineering, Kulliyyah of Engineering, International Islamic University Malaysia, 53100 Gombak, Malaysia

E. M. Nia et al. (eds.), *Sustainable Development Approaches*, Lecture Notes in Civil Engineering 243, https://doi.org/10.1007/978-3-030-99979-7_1

1 Introduction

The public bus passengers focused on several influencing factors including cost, freedom, travel time, comfort, availability, and reliability [1, 2, 3, 4]. Reliability is the most significant factor as the bus delay caused dissatisfaction among bus riders and caused them not to use public transport for their daily mode of transportation [5, 6]. When there is a delay for a bus, the consecutive bus under a similar bus provider will also be delayed [7]. Bus delays are closely related to the irregularity of bus arrival time [8].

Another factor that caused the delay in bus arrival time is traffic congestion due to the high volume of vehicles on road [9]. In Malaysia, it was found that traffic congestion occurs due to increasing car ownership primarily in urban areas [10]. The number of private vehicles in Malaysia is known as one of the highest among other countries [11]. The traffic congestion is mostly caused by private vehicles especially during peak hours [12]. Therefore, diligent encouragement on the usage of public transportation is required for every road user. As more road users use public transport, less volume of the vehicle on the road and eventually improve the bus arrival time. Besides public buses, the rail transit system is also one of the best options for transportation alternatives [13].

The biggest transportation hub in Malaysia is Kuala Lumpur Sentral Station which gathers almost all public facilities such as buses, taxis, and rail and accommodates up to 180,000 users daily [14]. Usually, a transportation hub will act as the main terminal for buses to start their route and most users gather there to move from one place to another. The immense development of infrastructure and transportation networks in the past ten years has brought an increase in road networks of more than 87,626 km [15]. The road networks aid to cater the vast development of infrastructure, an increasing number of vehicles including buses, and a growing population.

The most significant attribute to public bus performance is the on-time performance. The public transport services including bus do not follow exactly their original schedule and caused users to lost their trust in the services [16, 17]. In Transit Capacity and Quality Service Manual (TCQSM) Third Edition [18], on-time performance is measured in percentage and classified based on the quality of service (QOS). From the measured percentage, the on-time performance QOS is defined. It is an important aspect that should be addressed to operators and authorities in managing and improving public bus service [19]. Financial allocation for public bus QOS improvement also needs to be included which depends on the executive governmental functionaries [20].

This paper focuses on analysing the on-time performance of public bus service in Northern Peninsular Malaysia. The performance is measured by routes available at each capital city in the study area.

2 Methods

The states located in Northern Peninsular Malaysia are Perlis, Kedah, Pulau Pinang, and Perak. In this study, the capital cities of each northern state namely, Kangar, Alor Setar, Georgetown, and Ipoh is the coverage area of the study. The bus scheduled and arrival of each route were collected as the primary data that was then triangulated with secondary data in the form of literature reviews, manuals, and textbooks [21]. According to both deductive and inductive approaches, the analysis employed in this study considered thematic analysis [22].

2.1 *Bus Route*

Every bus provider has a different number of routes and coverage areas. Most of the initial point of the bus routes is at the main terminal of the city. Every bus at every route is tracked with their on-time schedule. The data is collected during the hot weather all the time. The bus operation starts as early as 5.00 am and as late as 10.00 pm.

For Kangar, six trunk routes and four feeder routes available that cover area of Kangar, Arau, Padang Besar, Ayer Hitam, Kuala Perlis, Felda Chuping, Beseri, Kilang Gula, Shabab Perdana, Universiti Utara Malaysia (UUM), Changlun, Jitra and Pekan Baru. The routes for myBAS Kangar taken account in analysing the on-time performance are T10 (Changlun by Arau), T11 (Padang Besar), T12 (Kuala Perlis), T13 (Alor Setar by Ayer Hitam), T14 (Alor Setar by Jitra), F100 (Seriab by Panggau), F101 (Seriab by Padang Behor), and F102 (Kilang Gula).

All routes in Alor Setar start at the main terminal of Shabab Perdana, Alor Setar. The routes include in this study T13 (Kangar to Ayer Hitam), T14 (Kangar to Jitra), T15 (Sintok (UUM)), Kuala Kedah, Kuala Nerang, Baling, Sungai Petani to Yan, Pokok Sena to Kuala Nerang, Langgar to Sungai Tiang, Bukit Raya to Sik, Tokai to Sungai Tiang, Kuala Ketil to Kulim and Sungai Petani.

There are more than 50 routes with 1998 bus stops in Georgetown. Some of the areas are Georgetown, Gelugor, Jelutong, Batu Lanchang, Bayan Baru, Sungai Nibong, and many more. The routes considered for Georgetown are 101 (Jetty to Teluk Bahang), 200 (Jetty to Penang State Mosque), 201 (Jetty to Paya Terubong by Air Itam), 202 (Jetty to Paya Terubong by Farlim), 203 (Jetty to Air Itam by Farlim), 204 (Jetty to Penang Hill Lower Station), 206 (Lotus's Penang to Jetty), CAT, 10 (Jetty to Botanic Gardens), 11 (Jetty to Jalan Tan Sri The Ewe Lim), 12 (Jetty to Bandar Sri Pinang), 301 (Jetty to Relau by Sungai Dua), 302 (Jetty to Batu Maung by Bukit Gambir), 303 (Jetty to Bukit Gedung), 401 (Jetty to Balik Pulau By Bayan Baru), 401E (Jetty to Balik Pulau by Tun Dr. Lim Chong Eu Expressway), and 502 (Pekan Genting to Jetty).

In Ipoh, the bus routes mostly connected to Medan Kidd Terminal and Amanjaya Terminal. The areas covered are Chemor, Tasek, Tanjong Rambutan, Medan

Gopeng Bus Station, Paser Puteh, Pengkalan Sentosa, Batu Gajah, Sri Iskandar, Taman Botani, Bercham, Manjoi, Buntong, Taman Cempaka and Ampang. The routes in Ipoh explored in this study are T30a (Medan Kidd to Amanjaya), T30b (Medan Kidd to Chemor by Amanjaya), T31a (Medan Kidd to Chemor), T31b (Medan Kidd to Chemor by Jalan Tun Abdul Razak), T32 (Medan Kidd to Bercham), T33a (Medan Kidd to Chemor by Tanjung Rambutan), T33b (Medan Kidd to Tanjung Rambutan), T34 (Medan Kidd to Gopeng), T35 (Medan Kidd to Pengkalan Sentosa), T36 (Medan Kidd to Seri Iskandar), T37 (Medan Kidd to Botani), F100 (Medan Kidd to Pekan Ipoh), F101a (Medan Kidd to Taman Ipoh), F101b (Medan Kidd to JPJ), F102 (Medan Kidd to Buntong), F103 (Medan Kidd to Ampang), 35 (Medan Kidd to Kuala Kangsar), 37 (Medan Kidd to Tg. Tualang), 39 (Medan Kidd to Bruas), 46 (Medan Kidd to Manjung), and 66 (Medan Kidd to Kampar).

2.2 On-Time Performance of Public Bus Transportation

Every people will face challenges that required them to achieve a certain degree of autonomous decision-making [23]. As for public bus users, the challenge faced by them is the bus scheduled time. The delay in bus arrival time led them to make fast decision-making of public transportation alternatives. The result from the on-time performance of a public bus will determine the quality of service (QOS). The same goes for other attributes such as passenger load factor, service frequency, and headway adherence. Referring to Transit Capacity and Quality Service Manual (TCQSM) Third Edition [18], on-time performance QOS is defined as the 0 to 5 min late whether on departure on the arrival of public bus transportation.

The on-time performance QOS for public bus transportation is analyzed by the on-time percentage shown in Eq. 1. High results of on-time performance percentage indicate more vehicles follow the assigned bus schedules. On-time performance is very important for public bus service as the users can plan their rides hassle-free. When the on-time performance percentage is determined, the percentage is rated based on their quality of service in Table 1 [18, 24]. The best and least QOS would be A and E respectively. Perspectives for passengers and operation on each QOS are explained in Table 1.

$$On-time\ performance\ percentage = \begin{pmatrix} Total\ on-time\ arrivals \\ \div Total\ actual\ arrivals \end{pmatrix} \times 100\,\% \quad (1)$$

Table 1 Quality of service for on-time performance as indicated in Transit Capacity and Quality Service Manual (TCQSM) Third Edition [18]

Quality of Service	On-time performance percentage (%)	Passenger perspective	Operator perspective
A	95–100	Passengers making one round trip per weekday with no transfer experience one, not on-time vehicle every 2 weeks	Achievable by transit services operating below capacity on a grade-separated guideway not shared with non—transit vehicles, with few infrastructures of vehicle problems
B	90–94	Passenger making one round trip per weekday with no transfers experiences one, not on-time vehicle every week	Not -Achievable by transit services operating on a grade-separated guideway not shared with non—transit vehicles
C	80–89	Passenger making one round trip per weekday with no transfers experiences one on-time vehicle every week	Typical range for commuter rail that shares tracks with freight rail Typical range for light rail with some street running Achievable by bus services in small to mid-sized cities
D	70–79	Passenger making one round trip per weekday with no transfers experiences up to three not-on-time vehicles every week Passengers making one round trip per weekday with a transfer experience on-time vehicle every day	Typical range for light rail with a majority of street running Achievable by bus services in large cities
E	< 70	Service is likely to be perceived as highly unreliable	May be the best possible result for mixed traffic operations in congested CBD's

3 Results and Discussion

There are a total of eight routes in Kangar being investigated in this study. The result showed in Fig. 1 that four routes (T10, T13, T14, F101) fall under QOS of C (80–89%) for bus on-time performance. According to TCQSM 3rd Edition, QOS C means the passenger used bus service at least one round trip during weekdays and experience only a one-time bus service that follow the schedule. But one route in Kangar showed a very good performance in departing at the scheduled time which is route T12 (Kuala Perlis) of QOS A (100%). While route T11 falls under QOS D and F100 and F102 fall under QOS E. All and all, the average percentage of the on-time bus performance in Kangar is 80% at QOS of C.

In Alor Setar, the on-time performance for 9 out of 13 bus routes showed less than 70% which is at the lowest QOS (QOS E) in Fig. 2. The routes include Kuala Kedah, Kuala Nerang, Sungai Petani to Yan, Pokok Sena to Kuala Nerang, Langgar to Sungai

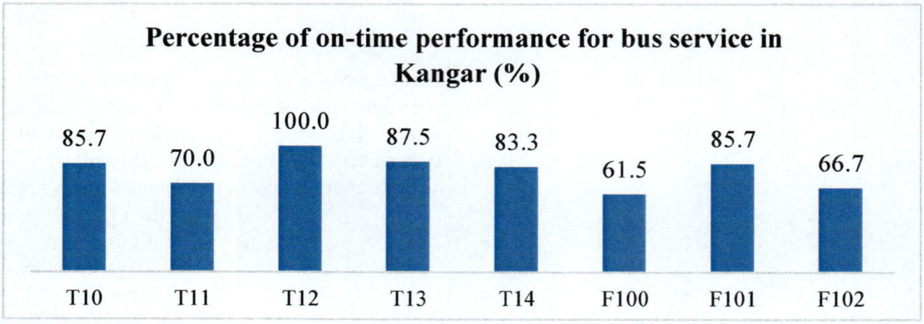

Fig. 1 Percentage of on-time performance for bus service in Kangar

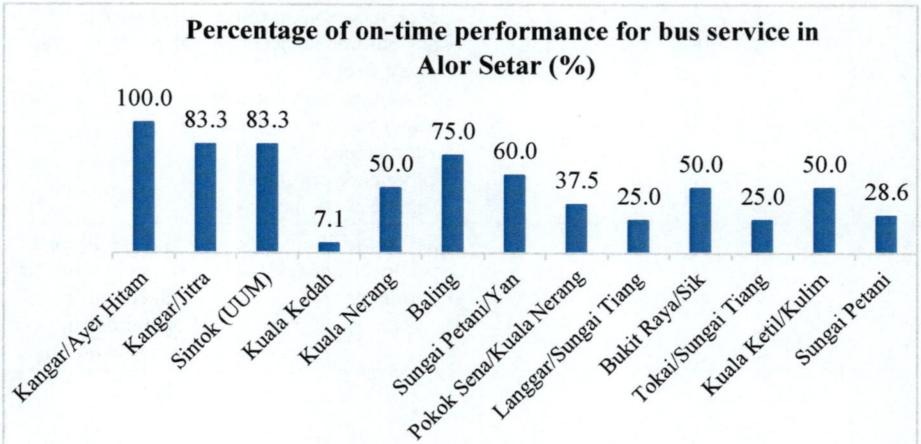

Fig. 2 Percentage of on-time performance for bus service in Alor Setar

Tiang, Bukit Raya to Sik, Tokai to Sungai Tiang, Kuala Ketil to Kulim and Sungai Petani. The bus service for the routes with QOS E is marked as a highly unreliable service for passengers and usually, the routes have mixed traffic congestion. The on-time performance for four more routes is at QOS A, C, and D. Bus route of T13 of Kangar to Ayer Hitam recorded at the maximum percentage of 100% for on-time performance. Bus route Kangar to Jitra (T14) and Sintok (UUM) (T15) is at QOS C with the same on-time performance percentage of 83.3%. The bus route for Baling is defined at QOS D with a 75.0% on-time performance percentage. Overall, the on-time performance percentage for bus routes in Alor Setar is 52% at QOS E which puts the bus service as undependable for passenger daily mode of transportation.

The on-time performance for bus service in Georgetown is illustrated in Fig. 3. All routes in Georgetown are found to be below 70% that fall directly in QOS E. The range of on-time percentage for bus routes in Georgetown is from 1.9 to 20.6%. The percentage is very low and highly unreliable from the passenger's perspective as established in TCQSM 3rd Edition. Although the available bus routes in Georgetown are wide and high frequency, the real-time performance is still weak may be due to bad traffic congestion. It can be concluded that the bus routes in Georgetown have an average on-time performance percentage of 9% with QOS E.

The percentage of on-time performance for bus service in Ipoh according to their routes is presented in Fig. 4. The on-time performance percentage for the routes vary with different grade of quality of service (QOS). Three routes of T32 (Bercham), F102 (Buntong), and F103 (Ampang) are classified as QOS A with the same calculated on-time performance percentage at 100%. These routes have perfect on-time performance following the scheduled bus arrival and departure time. As for routes T31b (Chemor by Jalan Tun Abdul Razak), T34 (Gopeng), and 35 (Kuala Kangsar), the on-time performance percentage is 93.8, 90.9, and 90.9% respectively which at QOS B.

Moreover, 7 out of 21 routes are classified as QOS C which are T30a (Terminal Amanjaya), T33a (Chemor by Tanjung Rambutan), T33b (Tanjung Rambutan), T35

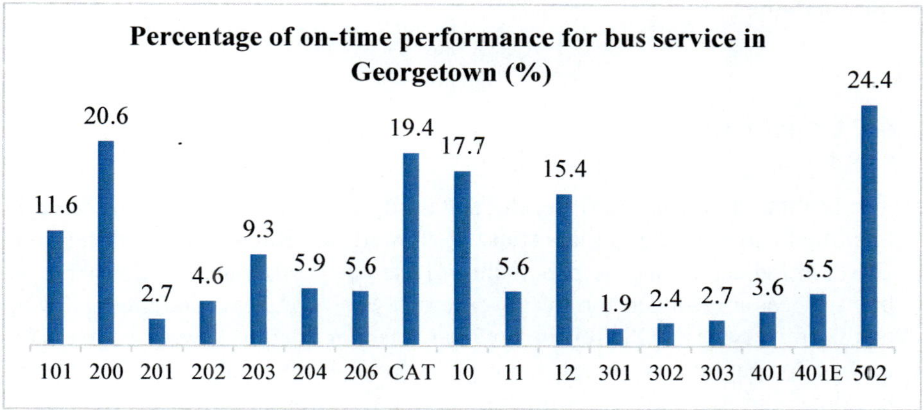

Fig. 3 Percentage of on-time performance for bus service in Georgetown

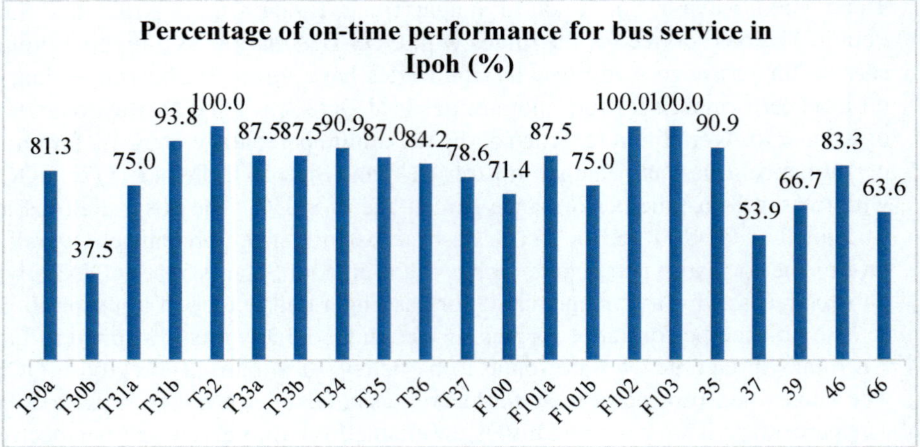

Fig. 4 Percentage of on-time performance for bus service in Ipoh

(Pangkalan Sentosa), T36 (Seri Iskandar), F101a (Taman Ipoh), and 46 (Manjung) with 80–89% on time performance percentage. Four routes of T31a (Chemor), T37 (Botani), F100 (Pekan Ipoh), and F101b (JPJ) fell into QOS D and another four routes of T30b (Chemor via Terminal Amanjaya), 37 (Tg. Tualang), 39 (Bruas), and 66 (Kampar) fell into QOS E. Hence, the average on-time performance percentage for bus service in Ipoh is 81%, classified as QOS C.

The on-time performance percentage for bus services in Kangar, Alor Setar, Georgetown, and Ipoh has been portrayed in Figs. 1, 2, 3, and 4 respectively. The results showed that Kangar and Ipoh are classified as QOS C, while Alor Setar and Georgetown are classified as QOS E. As established in TCQSM 3rd Edition, with QOS C, passengers make one round trip per weekday with at least one time the bus followed the scheduled time. As for QOS E, the bus service is considered unreliable and unlikely suitable for people that have fixed schedules for their daily use of transportation.

4 Conclusions

The on-time performance for each capital city in Northern Peninsular Malaysia is identified and evaluated in this study. It showed that Kangar and Ipoh fell in QOS C which indicated that the punctuality of the bus performance is debatable as the bus services are funded by the government. For Alor Setar and Georgetown, the quality of service (QOS) fell in E which was identified as a very weak quality of service for public bus transportation. The unreliability showed by the bus service in Alor Setar and Georgetown caused the passenger to opt for a different mode of transportation with better on-time performance. Therefore, the authorities and stakeholders need to improve the quality of service for public bus service not only in Alor Setar and Georgetown but also Kangar and Ipoh to gain back the passenger's trust in the punctuality of bus service of the respective cities. Also, to ensure the passenger still use the public bus as their daily mode of transportation. The evolution of trends and technology allows the researchers to have a vast number of scientific and technological resources, so the analysis of on-time performance for a public bus will be improved from time to time [25].

Acknowledgements Authors wishing to acknowledge Yayasan Canselor Universiti Tenaga Nasional for providing a research grant for this study (202101018YCU) and Innovation Research Management Centre (IRMC), Universiti Tenaga Nasional.

References

1. Olojede O, Yoade A, Olufemi B (2017) Determinants of walking as an active travel mode in a Nigerian city. J Transp Heal 6(February):327–334. https://doi.org/10.1016/j.jth.2017.06.008
2. Fadhlina S, Fisal M, Sabahiah N, Sukor A (2021) Level of Service (LOS) and bus headways: a case study of George Town, Penang. J. Kejuruter. 33(3):623–632
3. Jain S, Aggarwal P, Kumar P, Singhal S, Sharma P (2014) Identifying public preferences using multi-criteria decision making for assessing the shift of urban commuters from private to public transport: a case study of Delhi. Transp Res Part F Traffic Psychol Behav 24:60–70. https://doi.org/10.1016/j.trf.2014.03.007
4. De Luca S (2014) Public engagement in strategic transportation planning: an analytic hierarchy process based approach. Transp Policy 33:110–124. https://doi.org/10.1016/j.tranpol.2014.03.002
5. Fonzone A, Schmöcker JD, Liu R (2015) A model of bus bunching under reliability-based passenger arrival patterns. Transp Res Procedia 7:276–299. https://doi.org/10.1016/j.trpro.2015.06.015
6. Ishaq R, Cats O (2020) Designing bus rapid transit systems: lessons on service reliability and operations. Case Stud Transp Policy 8(3):946–953. https://doi.org/10.1016/j.cstp.2020.05.001
7. Bellei G, Gkoumas K (2010) Transit vehicles' headway distribution and service irregularity. Public Transp 2(4):269–289. https://doi.org/10.1007/s12469-010-0024-7
8. Wahab RA, Borhan MN, Rahmat RAAOK (2017) Prediction of bus arrival times at bus stop. Int J Technol 1:160–167. https://doi.org/10.1177/001452460608967
9. Chioni E, Iliopoulou C, Milioti C, Kepaptsoglou K (2020) Factors affecting bus bunching at the stop level: a geographically weighted regression approach. Int J Transp Sci Technol 9(3):207–217. https://doi.org/10.1016/j.ijtst.2020.04.001
10. Norhisham S et al (2019) Statistical overview on quality bus services in klang valley. Int J Adv Sci Technol 28(10):370–380
11. Kamba AN, Rahmat RAOK, Ismail A (2007) Why do people use their cars: a case study in Malaysia. J Soc Sci 3(3):117–122. https://doi.org/10.3844/jssp.2007.117.122
12. Norhisham S et al (2021) Development of expert system for bus services in Klang valley. IOP Conf Ser Earth Environ Sci 708(1). Doi: https://doi.org/10.1088/1755-1315/708/1/012039
13. Kasehyani NH, Abd Rahman N, Abdul Sukor NS, Halim H, Katman HY, Abustan MS (2019) Evaluation of pedestrian walking speed in rail transit terminal. Int J Integr Eng 11(9):26–36. Doi: https://doi.org/10.30880/ijie.2019.11.09.003

14. Kasehyani NH, Abd Rahman N, Abdulsukor NS, Halim H, Katman HY, Sabustan M (2018) A theoretical framework for understanding pedestrian behaviour attributes based on spatial interaction. IOP Conf Ser Mater Sci Eng 374(1). Doi: https://doi.org/10.1088/1757-899X/374/1/012089
15. Al-Saffar ZH et al (2021) A review on the durability of recycled asphalt mixtures embraced with rejuvenators. Sustain 13(16):1–24. https://doi.org/10.3390/su13168970
16. Abu Bakar MF, Norhisham S, Fai CM, Baharin NL (2021) Evaluating the quality of service for bus performance in Kuantan. Int J Acad Res Bus Soc Sci 11(2):1342–1351. Doi: https://doi.org/10.6007/ijarbss/v11-i2/9209
17. Suh W, Park S, Lee E (2011) Fault tolerant intelligent transportation systems with an agent. Commun Comput Inf Sci 195 CCIS:16–25. Doi: https://doi.org/10.1007/978-3-642-24267-0_3
18. Kittelson I, Assoc I (2013) Parsons Brinckerhoff, I. KFH Group, Texas A&M Transportation Institute, and Arup. Transit capacity and quality of service manual, 3rd Edition [Online]. http://www.trb.org/main/blurbs/169437.aspx%5Cn, http://www.worldtransitresearch.info/research/4941/
19. Norhisham S et al (2021) Evaluating the quality of services for bus performance in Alor Setars. IOP Conf Ser Earth Environ Sci 708(1). Doi: https://doi.org/10.1088/1755-1315/708/1/012038
20. Nor MZM, Mahdzir N, Mohamad AM (2020) Acceptance of conventional insurance principles as takaful basic principles: Shariah and legal analysis. J Crit Rev 7(8):1550–1553. https://doi.org/10.31838/jcr.07.08.306
21. Mohamad AM, Hamin Z, Md Nor MZ, Kamaruddin S, Nizam Md Radzi MS (2020) The implications of audio/video conference systems on the administration of justice at the Malaysian Courts. Webology 17(2):904–921. Doi: https://doi.org/10.14704/WEB/V17I2/WEB17076
22. Nor MZM, Mohamad AM, Azhar A, Latif HM, Khalid AHM, Yusof Y (2019) Legal challenges of Musharakah Mutanaqisah as an alternative for property financing in Malaysia. J Leg Ethical Regul Issues 22(3)
23. Mohamad AM, Salleh ASM, Nor MZM, Yusuff YMI (2020) Impacts of augmented reality in legal studies: students' reflections. 2020 Seventh International Conference on Information Technology Trends (ITT) 2020:151–155
24. Norhisham S, Katman HY, Ismail N, Abd Halim SNN, Ismail A, Borhan MN (2018) Evaluation on time performance for public bus service in West klang valley. Int J Eng Res Technol 11(3):403–415
25. Omar MF, Nawi MNM, Jamil JM, Mohamad AM, Kamaruddin S (2020) Research design of mobile based decision support for early flood warning system. Int J Interact Mob Technol 14(17):130–140. https://doi.org/10.3991/ijim.v14i17.16557

Methods on Calculating the International Roughness Index: A Literature Review

Juland A. Padilla and Orlean G. dela Cruz

Abstract The pavement management system determines the optimum maintenance techniques for each mile of the highway network. The total irregularities in the pavement surface per linear travel unit distance affect the ride quality. As a result, the safety of road users is known as pavement roughness or roughness index. The research aims to discover the various methods and procedures for computing IRI. It also wants to investigate how IRI affects the driver, the vehicle, and road conditions. The connection between road condition, roughness, and quality can impact the IRI values. Finally, considerations to various challenges and recommendations are in the proposal for evaluating IRI.

Keywords International Roughness Index · Pavement management system · Road conditions · Road roughness · Sensor-based monitoring system

1 Introduction

Highway structure should guarantee safe driving, the thoroughfare geometric design must be appropriate, and supporting facilities and management standards should be comparable [1]. Creating a pavement management system (PMS) determines the best maintenance strategies for each mile of the highway network. PMS uses pavement roughness measurement at the network and project levels [2, 3].

Pavement roughness is the sum of imperfections in the roadway surface per linear travel unit distance that degrade the ride quality of an automobile and, as a result, the safety of road users [3]. The assessment of the metrics for pavement roughness as ride quality determines the state of the pavement [4]. One of the specific measures is the International Roughness Index (IRI). The World Bank created IRI in the 1980s

J. A. Padilla
Civil Engineering Department, Technological Institute of the Philippines, Quezon City, Philippines
e-mail: japadilla.ce@tip.edu.ph

J. A. Padilla · O. G. dela Cruz (✉)
Graduate School, Polytechnic University of the Philippines, Manila, Philippines
e-mail: ogdelacruz@pup.edu.ph

E. M. Nia et al. (eds.), *Sustainable Development Approaches*, Lecture Notes in Civil Engineering 243, https://doi.org/10.1007/978-3-030-99979-7_2

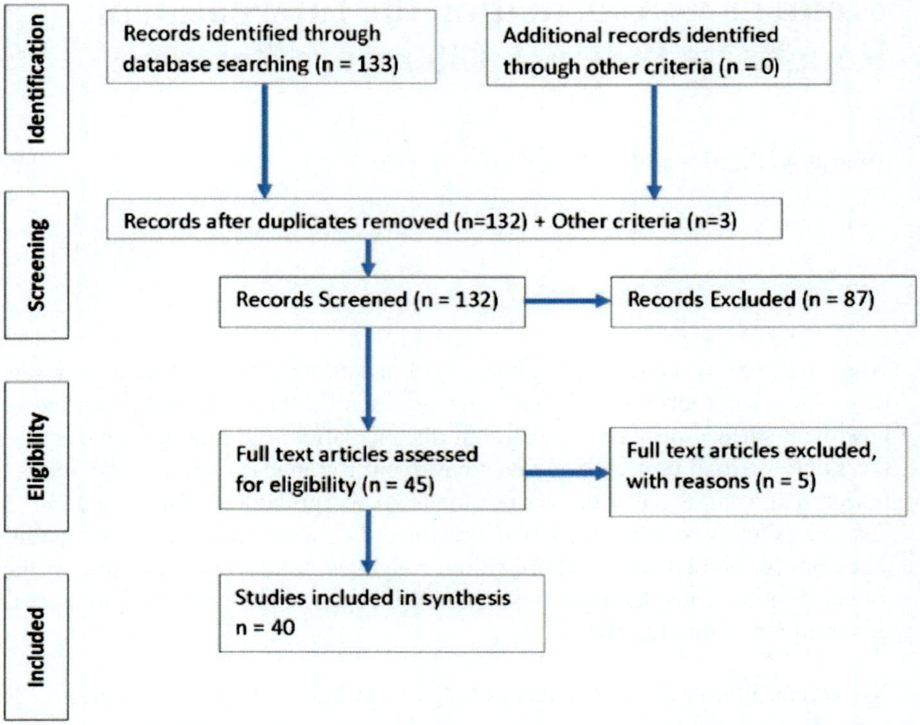

Fig. 1 Methodology of selecting past articles

for consistent infrastructure appraisal in emerging nations [5]. For highway maintenance and maintaining the safety and comfortability of road users or minimizing the dynamic load of cars and pavement, a precise characterization of the road roughness of longitudinal highway profile is required [6].

IRI claims that road surface roughness is generally acknowledged and used by road management authorities worldwide [7]. Moreover, IRI evolved from its establishment and renovated itself to cope with national needs. The study aims to determine the different methods and approaches in calculating IRI, connecting various aspects and observing some included or excluded factors for calculating IRI.

2 Methodology

This literature study aims to find empirical research examining the impacts and methods of calculating International Roughness Index values. To obtain topics to be discussed in the following, a comprehensive literature analysis to present data collecting using keywords such as the International Roughness Index. Scopus is one of the most widely utilized academic and practitioner journals search engines.

Using the Title/Abstract/Keyword in the database to search. One hundred thirty-three journals are selected and undergo screening to obtain four topics.

The Fig. 1 shows the selection process of past articles to create four topics.

3 Discussion

3.1 Methods in Getting IRI Values

The international roughness index (IRI) is a tool to identify the state of pavement degradation. The IRI value grows as pavement age increases [8]. The degree of surface roughness is one of the key factors supporting pavement degradation [9]. IRI computation system transforms longitudinal and vertical profile data into a vehicle motion response [10]. The road profile is measured using the dynamic response of cars and then simulated using the so-called quarter car method in which IRI is used in several countries [11] (Table 1).

3.1.1 Conventional Approach in Getting IRI Values

Data gathering is an essential element of the pavement control system's execution. Using automated systems to track vast road networks is accurate and saves time [1]. The Kinect depth data was validated and adjusted as the initial stage in this investigation. Using the ASTM E1926 method, the section has computed IRI value was determined logically and engineeringly by field observation and expert knowledge. It was also verified using a precise manual device [20].

Table 1 The approach of getting IRI Values

Study	Methods of getting IRI	Standard	Modern
[8, 12]	GMDH and ANN method	x	
[10, 11]	Quarter Car Model	x	
[1]	Road Roughness Monitoring	x	
[4]	Real-Time Fitness Analysis		x
[3, 13, 14]	Use of accelerometer of a smartphone		x
[15–17]	Using roadroid on a smartphone		x
[18]	Inverse Pseudo Excitation Method	x	
[19]	Physiological Pavement Evaluation	x	
[20]	Use of a Low-Cost Sensor	x	
[21, 22]	Using Regression Analysis	x	
[23]	Vehicle suspension car model	x	

Another IRI calculation is the Inverse Pseudo Excitation Method (IPEM). Experimented with speed, road roughness rating, type of vehicle, and tire enveloping characteristics influenced the new strategy [18]. In addition, regression analysis based on applying the theory of jerk equation can evaluate road roughness wherein jerk is the result of the sudden change in acceleration which affects riders comfortability [22]. Moreover, conducting a preliminary study of the physiological assessment of pavement roughness showed the influence of road roughness on passenger safety and comfortability [19].

3.1.2 Modern Approach in Getting IRI Values

A smartphone app can test real-time bitumen road and vehicle performance [4]. It found out that FFS and V50 have a significant association with road roughness. Furthermore, calibration of the road-induced acceleration values to produce the PSD of research paths using phones. Moreover, Mathlab Code investigates the relationship between the square root of the PSD of raw acceleration values and average IRI values [13]. In most cases, increasing IRI value decreases speed rapidly [14]. Under natural driving conditions, vehicle vibration response can be a robust measure of pavement roughness [3].

3.2 IRI and Safety

Road surface degradation and deterioration in road safety are of particular concern. [23]. Recent evidence shows the association between pavement roughness and life use in road surface management. The established International Roughness Index is indirectly proportional to the vehicle's prescribed safe speed [7]. Using the path M&R cost constraint index, the calculation for safety enhancement satisfies the limits on preservation and rehabilitation costs by applying the Floating Fixed-length Segmentation procedure to the chosen highway [24]. For freight transport, well-maintained roads are essential since they help reduce vehicle and cargo damage during transportation [25].

The Dynamic Load Coefficient determines the relationship of IRI to the safety of ride quality [6]. A real-world vehicle trial established the linear correlation between IRI and the driving workload. Aside from affecting ride quality and comfort, road surface roughness on the theory and calculation of driving workload and the comfort and safety threshold [26]. The safety results are relevant: drivers pay less attention to secondary triggers as the IRI rises [27]. The development of four safety efficiency functions to collect data on the safety implications of IRI-measured pavement surface roughness. The measurement uses single-vehicle accidents to develop the four safety efficiency aspects in a complete, rear-end, same-direction sideswipe [28].

3.3 IRI as Road Quality Indicator

The evaluation of road conditions is crucial in determining the extent of road injury [15]. To prioritize road maintenance, IRI is a commonly used parameter for quantifying road roughness and identifying the level of degradation for all roads. A low IRI value indicates a smooth lane, while a high value indicates a distressed road [29]. Prof. Vasilyev developed a dynamic approach for estimating highway maintenance efficiency that allows for a comprehensive and reasonable evaluation of road maintenance [30]. Various tools such as the Profilograph, Response-Type instruments, Walking Profilers, and Inertial Laser Profilers are available to test IRI [16].

Permanent deformation is a significant load-related discomfort in flexible pavement structures and worsens with load repetitions, impacting road roughness, serviceability, and the international roughness index (IRI) [31]. Roadroid tools extract IRI values by pairing SDI values to determine appropriate maintenance recommendations [17]. Because of the non-stationary nature of real road profiles, they vary significantly from artificial ones. Appropriate in situ measurements are needed to determine road pavement roughness and user impact [32]. Although certain distresses have a little discernible effect on maps, the a_{wz} method identifies essential parts for proper assessments and surveys that determine the distresses' causes to choose suitable maintenance measures [33].

A customer-oriented level of service is putting more pressure on high-quality road management. Obtaining roughness data in road networks for running PMS and using the IRI to assess public roads in local cities is essential for road administrators. STAMPER has profited substantially by permitting the computation of real-time road surface profiles without the need for a specific vehicle. It is also feasible to simultaneously collect road profiles for both wheel directions and IRI and acceleration measurements [34]. However, expansive soils will hasten pavements' degradation, resulting in increased cracking and roughness [35].

3.4 Effects of IRI in Road Conditions

Representing roughness indices in recent decades to signify pavement surface characteristics and relative performance [2]. Roughness on urban roadways is impacted heavily by the terrain's natural topography elements and features [36]. IRI and crash rate have a unique connection in all circumstances, suggesting that crash rate does not essentially rise to a particular IRI number [37]. Due to the vicious loop formed between the mutually aggravated roughness and the increase in dynamic vehicles/pavement contact forces, the growth in pavement roughness and its degradation increases in a non-linear manner [38]. The IRI measures the functional performance of road pavements and acts as a structural performance indicator [39]. However, measurement variability can affect IRI determinations by calculating the IRI value of a road segment on a single test run [40, 41]. Preliminary vehicle response/fatigue damage estimates require a thorough grasp of the road profile and its objective roughness categorization. The use of IRI and Power Spectral Density (PSD) to objectively classify a road's severity [5].

4 Challenges and Gaps

The IRI has become commonly used as a roughness index indicator for analyzing and maintaining road networks [5]. Based on the study, the relationship of road condition, roughness, and quality can affect the outcomes of the values generated for IRI [5, 11, 34, 37–39]. Hence, the load applied to each road specification is due to road and vehicle conditions disputes such as damaged roads [4, 14, 24, 26, 35]. Moreover, even a slight change in vehicle speed and vehicles can also affect the values [2, 3, 7, 15, 19, 28, 29, 33]. Both parts are likely to be evaluated as fair erroneously due to measurement uncertainty [39] given the many variables that influence road pavement roughness and evaluation, the utility of generated fake profiles for analyzing specific parameters and evaluating their effects on ride quality [27, 32, 41].

Although some research shows great promise and can disrupt specific established approaches, there is still space for improvement [1, 8, 18, 25, 34]. Even the versatility of the smartphone-based roughness assessment approach opens a world of options for improving pavement management decision-making. However, it may vary on the models used during the experiment [13, 16, 17]. Hence, errors in measurement procedures are unavoidable. These mistakes frequently add to the anticipated value modeling error, increasing the total error [8, 10, 23, 30, 31]. For any reason, the suggested approach for reconstructing a 3D model of a pavement surface is time-consuming and computationally intensive [20] (Table 2).

Table 2 Challenges considered in IRI Evaluation

Citation	Challenges
[5, 11, 34, 37–39]	Relationship of Road roughness, ride quality and road condition
[2, 3, 7, 15, 19, 28, 29, 33]	Type of Vehicle and Speed
[6, 25]	Road roughness measurement using a sensor-based monitoring system
[13, 16, 17]	Data collected varies on the cell phone model, vehicle, and speed
[4, 14, 24, 26, 35]	Road Condition
[8, 10, 23, 30, 31]	Creating another method to satisfy the result
[1, 8, 18, 25, 34]	Further analysis of other factors that may affect the result
[27, 32, 41]	Parameters Considered
[20]	Time-consuming in generating 3D model

When selecting road sections, the following characteristics may lead to horizontal bends with a minimum radius of 130 m. More damage roads to minimize and eradicate any impact on automobiles applying brakes at curves (60 km/h). Even minor changes in vehicle speed and vehicle type might impact the results. The International Roughness Index (IRI) calculation can vary in several ways. Some nations may even change these indexes to meet their national requirements while maintaining the universal idea. It is necessary to demonstrate that road conditions, such as degradation and driving behavior, can aid future studies.

5 Conclusions

Since 1980, the International Roughness Index (IRI) has a variety of calculation methods. The conventional and modern approach introduces many equations, methodologies, or techniques to prove the roughness index's accuracy. Some countries may even modify these indices to cope with their national standards while anchoring its principle to the international. However, some parameters and event data may be restricted during experiments to produce an outcome; moreover, it excludes some factors that may affect IRI results. The summary of the past research focuses on the systemic and empirical approach for calculating IRI. Hence, it is needed to show that road conditions, such as deterioration and drivers' behavior, may contribute to future research as it dives more profound into the analysis. Some research also recommends the consideration of other indices as to IRI. Moreover, further developments for the equipment's utilization for future study may also enhance the present equipment and modify it to accommodate the other factors.

Acknowledgements The researchers would like to thank ES Family, namely; Christian A. Mendoza, Kristel D. Lopez, Jobelle S. Dajac, John Pual J. Pauya, Ely D. Biago, and Armando N. Victoria Jr, for their continuous support in conducting this paper.

References

1. Arbabpour Bidgoli M, Golroo A, Sheikhzadeh Nadjar H, Ghelmani Rashidabad A, Ganji MR (2019, August) Road roughness measurement using a cost-effective sensor-based monitoring system. Autom Constr, vol 104. Doi: https://doi.org/10.1016/j.autcon.2019.04.007
2. Loprencipe G, Zoccali P, Cantisani G (2019, April) Effects of vehicular speed on the assessment of pavement road roughness. Appl Sci 9(9). Doi: https://doi.org/10.3390/app9091783
3. Zeng H, Park H, Smith BL, Parkany E (2018, August) Feasibility assessment of a smartphone-based application to estimate road roughness. KSCE J Civ Eng 22(8). Doi: https://doi.org/10.1007/s12205-017-1008-9
4. Venkatesulu S, Sudarshan E, Korra SN, Raghava Kumari D, Yadav BP, Mahender K (2020, December) Real time fitness analysis of Bitumen Road and vehicle through their acoustic signals. IOP Conf Ser: Mater Sci Eng 981(3). Doi: https://doi.org/10.1088/1757-899X/981/3/032004

5. Pawar PR, Mathew AT, Saraf MR (2018) IRI (International Roughness Index): an indicator of vehicle response. Mater Today: Proc 5(5). Doi: https://doi.org/10.1016/j.matpr.2018.02.143
6. Múčka P (2017, June) Road roughness limit values based on measured vehicle vibration. J Infrastruct Syst 23(2). Doi: https://doi.org/10.1061/(ASCE)IS.1943-555X.0000325
7. Nguyen X, Nguyen T, Hoa Tran P (2020, July) The effect of road surface roughness to recommended speed of vehicles. IOP Conf Ser: Mater Sci Eng, vol 886. Doi: https://doi.org/10.1088/1757-899X/886/1/012014
8. Ziari H, Sobhani J, Ayoubinejad J, Hartmann T (2015) Prediction of IRI in short and long terms for flexible pavements: ANN and GMDH methods. Int J Pavement Eng 17(9):776–788. https://doi.org/10.1080/10298436.2015.1019498
9. Savnns MW On the calculation of international roughness index from longitudinal road profile
10. Nurhadiansyah R, Hadiana A (2019, November) Toll road roughness index forecasting with combination grey forecasting model and similarity spatial data. IOP Conf Ser: Mater Sci Eng, vol 662. Doi: https://doi.org/10.1088/1757-899X/662/2/022065
11. Chen SL, Lin CH, Tang CW, Chu LP, Cheng CK (2020, December) Research on the international roughness index threshold of road rehabilitation in metropolitan areas: a case study in Taipei city. Sustainability (Switzerland) 12(24):1–19. https://doi.org/10.3390/su122410536
12. Ziari H, Sobhani J, Ayoubinejad J, Hartmann T (2016, October) Prediction of IRI in short and long terms for flexible pavements: ANN and GMDH methods. Int J Pavement Eng 17(9). Doi: https://doi.org/10.1080/10298436.2015.1019498
13. Janani L, Sunitha V, Mathew S (2020, January) Influence of surface distresses on smartphone-based pavement roughness evaluation. Int J Pavement Eng. https://doi.org/10.1080/10298436.2020.1714045
14. Abeygunawardhana C, Sandamal RMK, Pasindu HR (2020, July) Identification of the impact on road roughness on speed patterns for different roadway segments. Doi: https://doi.org/10.1109/MERCon50084.2020.9185387
15. Achmadi F, Suprapto M, Setyawan A (2017, February) The Priority of Road Rehabilitation in Karanganyar Regency Using IRI Estimation from Roadroid. IOP Conf Ser: Mater Sci Eng, vol 176. Doi: https://doi.org/10.1088/1757-899X/176/1/012033
16. Hossain MI, Tutumluer E, Nikita, Grimm C (2019, August) Evaluation of android-based cell phone applications to measure international roughness index of rural roads. Doi: https://doi.org/10.1061/9780784482575.034
17. Arianto T, Suprapto M, and Syafi'I (2018, March) Pavement condition assessment using IRI from roadroid and surface distress index method on national road in sumenep regency. IOP Conf Ser: Mater Sci Eng, vol 333. Doi: https://doi.org/10.1088/1757-899X/333/1/012091
18. Li J, Zhang Z, Wang W (2019, March) New Approach for Estimating International Roughness Index Based on the Inverse Pseudo Excitation Method. J Transp Eng, Part B: Pavements 145(1). Doi: https://doi.org/10.1061/JPEODX.0000093
19. Zhang C et al (2019, July) Study on the applicability of physiological method for evaluating pavement roughness. Doi: https://doi.org/10.1061/9780784482292.081
20. Khalifeh V, Golroo A, Ovaici K (2018, July) Application of an inexpensive sensor in calculating the international roughness index. J Comput Civ Eng 32(4). Doi: https://doi.org/10.1061/(ASCE)CP.1943-5487.0000761
21. Padilla JA, Victoria AN, dela Cruz OG, Despabeladera CT, Creencia CJN Evaluation of international roughness index by speed-related quality criteria in the Philippines. Proc Annu Int Conf Arch Civ Eng, pp 160–164. Doi: https://doi.org/10.5176/2301-394X_ACE19.523
22. dela Cruz OG, Mendoza CA, Lopez KD (2021, July) International roughness index as road performance indicator: a literature review. IOP Conf Ser: Earth Environ Sci 822(1):012016. Doi: https://doi.org/10.1088/1755-1315/822/1/012016
23. Bridgelall R (2014, March) A participatory sensing approach to characterize ride quality. Doi: https://doi.org/10.1117/12.2046854
24. Semnarshad M, Elyasi M, Saffarzadeh M, Saffarzadeh A (2018) Identification and prioritization of accident-prone segments using international roughness index identification and prioritization of accident-prone segments using …,"

25. Wessels I, Steyn WJvdM (2020, March) Continuous, response-based road roughness measurements utilising data harvested from telematics device sensors. Int J Pavement Eng 21(4). Doi: https://doi.org/10.1080/10298436.2018.1483505
26. Hu J, Gao X, Wang R, Sun S (2017) Research on Comfort and safety threshold of pavement roughness. Transp Res Rec 2641(1):149–155. https://doi.org/10.3141/2641-17
27. Kawamura A, Tomiyama K, Rossi R, Gastaldi M, Mulatti C (2017) Driving on rough surface requires care and attention. Transp Res Procedia, vol 22. Doi: https://doi.org/10.1016/j.trpro.2017.03.008
28. Lee J, Abdel-Aty M, Nyame-Baafi E (2020, February) Investigating the Effects of Pavement Roughness on Freeway Safety using Data from Five States. Transp Res Rec: J Transp Res Board 2674(2). Doi: https://doi.org/10.1177/0361198120905834
29. Zhao Y, Wang ML (2015, June) Measurement through dynamic tire pressure sensor inside the tire. Doi: https://doi.org/10.1061/9780784479216.026
30. Zhustareva Ev, Bochkarev VI (2020, June) The complex method of estimation of highway maintenance quality taking into account the International Roughness Index. IOP Conf Ser: Mater Sci Eng, vol 832. Doi: https://doi.org/10.1088/1757-899X/832/1/012035
31. Ghasemi P, Aslani M, Rollins DK, Christopher Williams R, Schaefer VR (2018, January) Modeling rutting susceptibility of asphalt pavement using principal component pseudo inputs in regression and neural networks. Int J Pavement Res Technol. Doi: https://doi.org/10.1016/j.ijprt.2018.01.003
32. Loprencipe G, Zoccali P (2017, March) Use of generated artificial road profiles in road roughness evaluation. J Mod Transp 25(1). Doi: https://doi.org/10.1007/s40534-017-0122-1
33. Loprencipe G, Zoccali P (2017, April) Ride quality due to road surface irregularities: comparison of different methods applied on a set of real road profiles. Coatings 7(5). Doi: https://doi.org/10.3390/coatings7050059
34. Abulizi N, Kawamura A, Tomiyama K, Fujita S (2016, October) Measuring and evaluating of road roughness conditions with a compact road profiler and ArcGIS. J Traffic Transp Eng (English Edition) 3(5). Doi: https://doi.org/10.1016/j.jtte.2016.09.004
35. Evans RP, Arulrajah A, Horpibulsuk S (2015, December) Detecting gilgai relief beneath sealed flexible pavements using road profile and road roughness measurements. Indian Geotech J 45(4). Doi: https://doi.org/10.1007/s40098-015-0164-4
36. Abudinen D, Fuentes LG, Carvajal Muñoz JS (2017, January) Travel quality assessment of urban roads based on international roughness index: case study in Colombia. Transp Res Rec: J Transp Res Board 2612(1). Doi: https://doi.org/10.3141/2612-01
37. Mamlouk M, Vinayakamurthy M, Underwood BS, Kaloush KE (2018, December) Effects of the international roughness index and rut depth on crash rates. Transp Res Rec: J Transp Res Board 2672(40). Doi: https://doi.org/10.1177/0361198118781137
38. Radović N, Jokanović I, Matić B, Šešlija M (2016, June) A measurement of roughness as indicator of road network condition – case study Serbia. *Teh Vjesn-Tech Gaz* 23(3). Doi: https://doi.org/10.17559/TV-20150212204747
39. Hassan R, Mcmanus K, Holden J (1999, January) Predicting Pavement deterioration modes using waveband analysis. Transp Res Rec: J Transp Res Board 1652(1). Doi: https://doi.org/10.3141/1652-57
40. Lu P, Tolliver D (2012, November) Pavement treatment short-term effectiveness in IRI change using long-term pavement program data. J Transp Eng 138(11). Doi: https://doi.org/10.1061/(ASCE)TE.1943-5436.0000446
41. Jia X, Huang B, Zhu D, Dong Q, Woods M (2018, June) Influence of measurement variability of international roughness index on uncertainty of network-level pavement evaluation. *J Transp Eng, Part B: Pavements* 144(2). Doi: https://doi.org/10.1061/JPEODX.0000034

Determining of Passenger Load Factor for Public Bus Transportation in Northern Peninsular Malaysia

Shuhairy Norhisham, Nor Najwa Irina Mohd Azlan, Muhammad Fadhlullah Abu Bakar, Noorazizun Mohd Saad, Wan Ahmad Faiz Wan Mohd Fauzi, Siti Aliyyah Masjuki, Mohd Zakwan Ramli, and Sarah Shaziah Samsudin

Abstract Over the decade, Malaysia showed excellent improvement in upgrading social mobility as the country moving to become a developing nation. The important factor of social mobility that needs to be highlighted is transportation. Day-to-day activities require transportation to move from one place to another. Having systematic and reliable public transportation ensures the well-being and satisfaction of users. This paper proposes to determine the passenger load factor specifically for public bus transportation in Northern Peninsular Malaysia. The number of passengers and the number of seats in a bus is needed to rate the passenger load factor according to their Quality of Service (QOS) from Transit Capacity and Quality Service Manual (TCQSM) 3rd Edition. The results of passenger load factor for Kangar, Alor Setar, Georgetown, and Ipoh in Northern Peninsular Malaysia were found to be at 0.07, 0.23, 0.56, and 0.14 respectively. The capital cities investigated in this study are rated as A for their QOS except for Georgetown at B. Although the QOS rated as excellent, some aspects need to be improvised such as better on-time performance and longer duration of service hours to encourage more users of public bus transportation and eventually increase the passenger load factor.

Keywords Public bus · Northern Peninsular Malaysia · Passenger load · Quality of Service · Public transportation

---

S. Norhisham (✉) · N. N. I. M. Azlan · M. F. A. Bakar · W. A. F. W. M. Fauzi · M. Z. Ramli · S. S. Samsudin
Department of Civil Engineering, College of Engineering, Universiti Tenaga Nasional, 43000 Kajang, Selangor, Malaysia
e-mail: shuhairy@uniten.edu.my

S. Norhisham · N. M. Saad · M. Z. Ramli
Institute of Energy Infrastructures (IEI), Universiti Tenaga Nasional, 43000 Kajang, Selangor, Malaysia

N. M. Saad
College of Computing & Informatics, Universiti Tenaga Nasional, 43000 Kajang, Selangor, Malaysia

S. A. Masjuki
Department of Civil Engineering, Kulliyyah of Engineering, International Islamic University Malaysia, 53100 Gombak, Malaysia

E. M. Nia et al. (eds.), *Sustainable Development Approaches*, Lecture Notes in Civil Engineering 243, https://doi.org/10.1007/978-3-030-99979-7_3

1 Introduction

Nowadays, speed is no longer a primary factor for the transportation system but, the comfort, security, and safety of the transportation is more important [1]. The development of a transportation system in a city considers the provision of infrastructure, private transportation, and public transportation. So, the role of the government is significant in improving the public transportation system of a city [1, 2]. The public transportation system is highlighted as a transformation to provide greater sustainability [3] because it can reduce traffic congestion thus, reducing the accumulation of traffic pollutants in the atmosphere [4]. Public transportation benefits the transportation sector by lowering air pollution to the well-being of the environment [5–7].

Among all the public transportation services available, buses receive the most riders which made up to 600,000 trips each day [8, 9]. Over the years, authorities and bus providers have made a lot of improvements on the service quality of bus including the management and monitoring system [10]. Transport planners also agreed that sustainable development of public transportation can be accomplished by providing the highest service quality [11]. Good quality of service is measured by customers' perception and demand [12]. Norhisham et al. reviewed four bus transit manuals of four different countries and experts suggested six attributes regarding the service quality of bus including service hours, passenger load factor, comparison car and bus travel, frequency of bus, on-time performance, and coverage of service area [13].

While research by Shen and Feng explained that two categories influence passenger's comfort when traveling by bus namely, vehicle facilities (related to comfortable seating) and passenger load factor (related to the frequency of buses) [14]. The influence factors for public bus quality of service arise from the problems faced by the passengers [15]. In a sustainable view, a high passenger load factor can lower down the emission gas per passenger. There are various studies of public transport passenger load factor because it is a notable indicator to bus service quality, useful for current and new users, relatively crucial factor for customer's satisfaction [1, 11, 16, 17]

According to Kittelson and Associates and KFH Group et al., the definition of passenger load factor is a ratio of actual bus riders to the number of seats on the bus [18]. The number of passenger load factor and ratio of bus occupancy is strongly connected from one another for the regular users on existing bus routes [19]. Low bus occupancy and passenger load factor may cause by low service quality and safety concerns [20].

This study focuses on determining the passenger load factor for public bus transportation in Northern Peninsular Malaysia. The passenger load factors of the cities in Northern Peninsular Malaysia will be rated according to their quality of service (QOS) of Transit Capacity and Quality Service Manual (TCQSM) Third Edition [18].

2 Methods

Determination of passenger load factor for Northern Peninsular Malaysia involved capital cities of Kangar, Alor Setar, Georgetown, and Ipoh. The parameters used to determine passenger load factors are the number of passengers and the number of seats in a route of the public bus. So, the number of passengers and number of seats in the public bus of each route is the primary data that will be bind together with secondary data in the manual of Transit Capacity and Quality Service Manual (TCQSM) Third Edition [21]. The analysis implied in this study considers deductive and inductive approaches which are considered as a conceptual analysis of passenger load factor [22]. To collect the data of the parameters, the steps taken are:-

1. Get ready at the bus terminal of the city.
2. Once a bus arrived at the starting point of the route, record the number of passengers and seats in the bus.
3. Then, at each stop of the route, record the number of passengers that come in and out of the bus.
4. Keep recording the number of passengers and ride the bus until the last stop.
5. Repeat step 1 until 4 for the return route or different route of the bus.

Once all the data is already being collected, the passenger load factor can be calculated using Eq. 1. Then, the passenger load factor is analyzed based on their respective quality of service. The quality of service (QOS) is rated based on the range and comment indicated from Transit Capacity and Quality Service Manual (TCQSM) 3rd Edition [18] as shown in Table 1. The rate of QOS is chosen based on different ranges of passenger load factors. For QOS A, B, C, D, and E, the range of passenger load factor are 0.00–0.50, 0.51–0.75, 0.76–1.00, 1.01–1.25, and 1.26–1.50 respectively. Only four routes being considered for each city because the passenger load factor and QOS are determined as an average.

Table 1 Quality of service for passenger load factor as indicated in Transit Capacity and Quality Service Manual (TCQSM) Third Edition [18]

Quality of Service	Passenger Load Factor (passenger/seat)	Comments
A	0.00–0.50	No passenger needs to sit next to another
B	0.51–0.75	Passenger can choose where to sit
C	0.76–1.00	All passengers can seat
D	1.01–1.25	Comfortable standee load for design
E	1.26–1.50	Maximum schedule load

$$\text{Passenger Load Factor} = \text{Number of Passengers} \div \text{Number of Seats} \quad (1)$$

3 Results and Discussion

A total of four bus routes are considered in calculating appropriate Quality of Service (QOS) based on their passenger load factor. The results of passenger load factor for bus routes at four individual capital cities are illustrated in Fig. 1 until Fig. 4. In Kangar, the passenger load factor for route T10 (Changlun by Arau), T11 (Arau), T12 (Kuala Perlis), and T100 (Seriab by Penggau) are 0.06, 0.09, 0.08, and 0.04 respectively. All route is at QOS A in range 0.00 to 0.50. The average passenger load factor for four routes in Kangar is 0.07, classified as QOS A. Bus riders in Kangar do not face any shortcomings in finding a seat when riding a bus as it has a low passenger load. It is also proven by Sani et al. [23] that citizens in Perlis choose to move from one place to another by motor instead of public transport because the routes are very limited.

The passenger load factor in Alor Setar is graphically shown in Fig. 2 of four routes namely, Kangar to Jitra, Sintok (UUM), Kuala Kedah, and Kuala Nerang. The results of the calculated passenger load factor for the route of Kangar to Jitra, Sintok (UUM), Kuala Kedah, and Kuala Nerang are 0.26, 0.29, 0.18, and 0.17 respectively. The results of passenger load factor for the bus routes in Alor Setar are below 0.50 which is at QOS A as established in TCQSM 3rd Edition. Hence, no bus riders are required to sit next to each other.

The passenger load factor in Georgetown is different from the other two capital cities of Kangar and Alor Setar because the range of passenger load factor varies. Bus routes 11 (Batu Lanchang), 101 (Tok Bahang), 204 (Bukit Bendera), and CAT (Weld Quay) recorded their passenger load factor at 0.27, 0.49, 0.55, and 0.92 respectively.

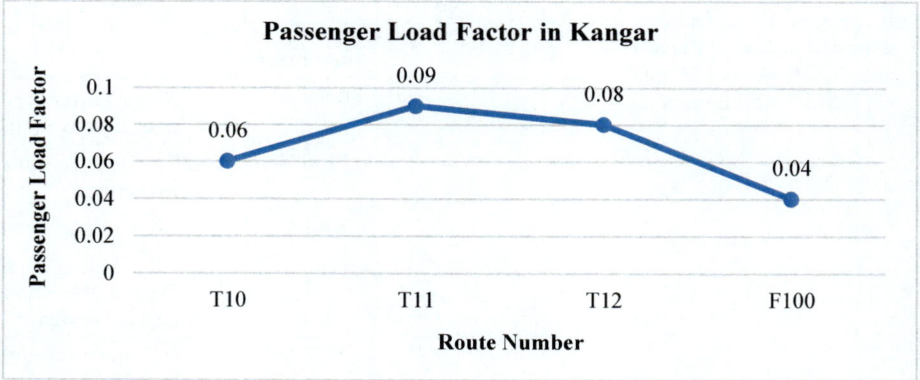

Fig. 1 Passenger load factor for bus service in Kangar

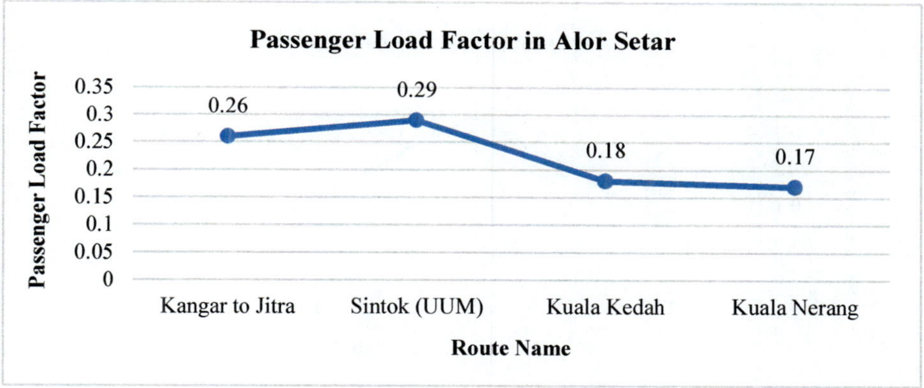

Fig. 2 Passenger load factor for bus service in Alor Setar

The bus routes of 11 and 101 are under QOS A, while bus routes of 204 and CAT are under QOS B and C respectively. The average passenger load factor for the bus routes in Georgetown is 0.56 which is classified as QOS B. According to TCQSM 3rd edition, the passenger load factor for public bus service at Georgetown indicates that passengers are free to choose their seat when riding the public bus either beside other riders or not upon availability (Fig. 3).

As for Ipoh, the passenger load factor for the bus routes is shown in Fig. 4. Route T30b (Chemor), T34 (Gopeng), T37 (Bandar Sri Botani), and F103 (Ampang) recorded a passenger load factor of 0.08, 0.22, 0.09, and 0.15 respectively which are in QOS A. Meanwhile for overall passenger load factor in Ipoh is 0.14 at QOS A. Based on the manual of TCQSM 3rd edition, the passengers can select any seat they want when riding a public bus in Ipoh due to the low passenger load factor. During the research, the state of Perak was having up to 30 new cases of Covid-19 daily hence, the anticipation of bus riders is declining. The citizens avoid public transport

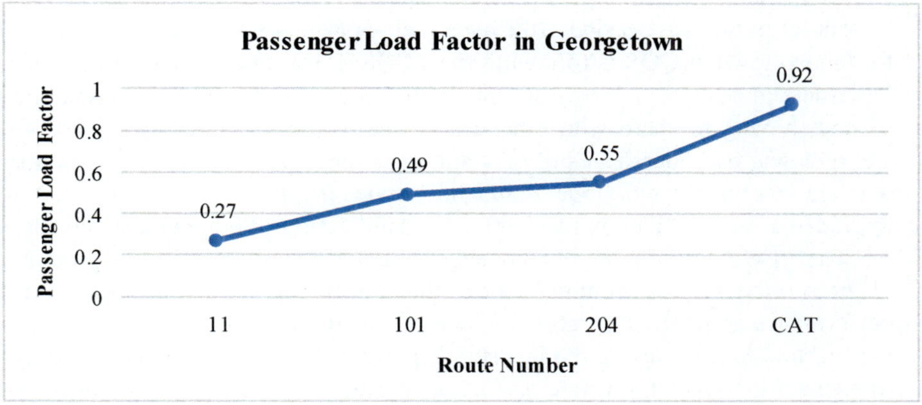

Fig. 3 Passenger load factor for bus service in Georgetown

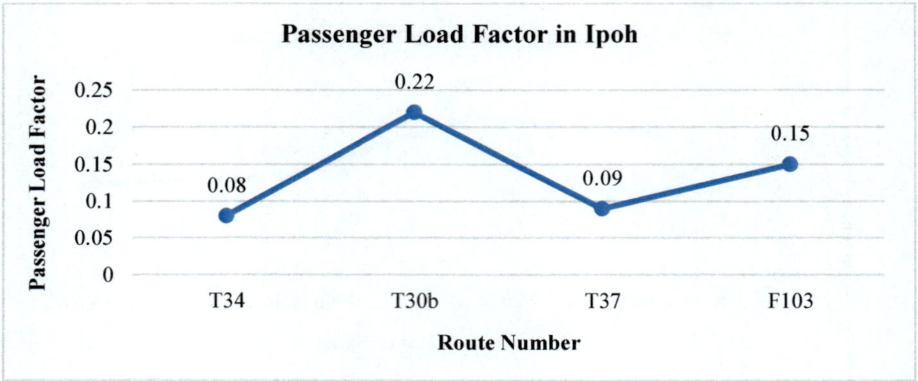

Fig. 4 Passenger load factor for bus service in Ipoh

especially with a high volume of passengers because the social distancing becomes invalid.

The quality of service (QOS) for public bus service per the passenger load factor at four of the capital cities in Northern Peninsular Malaysia were found to be at QOS A except for Georgetown at QOS B. The research was done during pandemic Covid-19 hence the rate of QOS may be different if the pandemic is under control. Apart from that, some state in Northern Peninsular Malaysia has small coverage of bus routes due to the size of the state, for example, Perlis. Therefore, citizens tend to choose private transportation rather than public transportation.

4 Conclusions

All and all, the average passenger load factor for Kangar, Alor Setar, Georgetown, and Ipoh were found at 0.07, 0.23, 0.56, and 0.14 respectively. The passenger load factors for public bus service at the capital cities were established as QOS A except for Georgetown at QOS B following the TCQSM 3rd Edition. This means that the bus riders at Kangar, Alor Setar, and Ipoh can select any seats in the public bus without having to share with other riders due to the low passenger load. As for Georgetown, bus riders can still pick any seats they want but the seat can either be with another bus rider or alone. Public bus service in all the capital cities needs to be upgraded in terms of their passenger load and utilization, although most of it has been rated as QOS A. Improving the passenger load factor will increase the productivity of the service. Other recommendations that can be highlighted are better on-time performance and higher duration of service which are closely related to passenger load factor. An increase in the bus rider's participation is influenced by the factors (passenger load factor, on-time performance, and service frequency). This research is done during the pandemic of Covid-19 so the results may be different when the pandemic is under control. Further research on the passenger load factor can be done in the situation of endemic Covid-19 to see the difference in the outcome of the study. Changes in trends and technology give opportunities to the researchers to have unlimited resources, hence the interpretation of passenger load factor for the public bus will be upgraded over time [24].

Acknowledgements Authors wishing to acknowledge Yayasan Canselor Universiti Tenaga Nasional for providing a research grant for this study (202101015YCU) and Innovation Research Management Centre (IRMC), Universiti Tenaga Nasional.

References

1. Pahala Y et al (2021) The influence of load factor, headway, and travel time on total fleet requirements and its implications for public transportation maintenance management on Transjakarta. Rev Int Geogr Educ 11(5):3422–3436. https://doi.org/10.48047/rigeo.11.05.231
2. Desertot M, Lecomte S, Gransart C, Delot T (2012) Intelligent transportation systems. Comput Sci Ambient Intell, pp 285–304
3. Hellekes J, Winkler C (2021) Incorporating passenger load in public transport systems and its implementation in nationwide models. Procedia Comput Sci 184:115–122. https://doi.org/10.1016/j.procs.2021.03.022
4. Pan Y, Qiao F, Tang K, Chen S, Ukkusuri SV (2020) Understanding and estimating the carbon dioxide emissions for urban buses at different road locations: a comparison between new-energy buses and conventional diesel buses. Sci Total Environ, vol 703. https://doi.org/10.1016/j.scitotenv.2019.135533
5. Jiaqiang E et al (2020) Heat dissipation investigation of the power lithium-ion battery module based on orthogonal experiment design and fuzzy grey relation analysis. Energy 211:118596. Doi: https://doi.org/10.1016/j.energy.2020.118596
6. Holland SP, Mansur ET, Muller NZ, Yates AJ (2021) The environmental benefits of transportation electrification: Urban buses. Energy Policy, vol 148. Doi: https://doi.org/10.1016/j.enpol.2020.111921
7. Sun L et al (2021) Reducing energy consumption and pollution in the urban transportation sector: a review of policies and regulations in Beijing. J Clean Prod, vol 285. https://doi.org/10.1016/j.jclepro.2020.125339
8. Chuen OC, Karim MR, Yusoff S (2014) Mode choice between private and public transport in Klang Valley, Malaysia. Sci World J no. Figure 1:7–9. Doi: https://doi.org/10.1155/2014/394587
9. Norhisham S et al (2020) Evaluating passenger load factor of public bus services in West Klang Valley. Lect Notes Civ Eng, pp 95–102
10. Abu Bakar MF, Norhisham S, Fai CM, Baharin NL (2021) Evaluating the quality of service for bus performance in Kuantan. Int J Acad Res Bus Soc Sci 11(2):1342–1351. Doi: https://doi.org/10.6007/ijarbss/v11-i2/9209
11. de Oña R, Eboli L, Mazzulla G (2014) Monitoring changes in transit service quality over time. Procedia-Soc Behav Sci 111:974–983. https://doi.org/10.1016/j.sbspro.2014.01.132
12. Azadi M, Shabani A, Khodakarami M, Farzipoor Saen R (2014) Planning in feasible region by two-stage target-setting DEA methods: an application in green supply chain management of public transportation service providers. Transp Res Part E Logist Transp Rev 7(1):324–338. Doi: https://doi.org/10.1016/j.tre.2014.07.009
13. Norhisham S, Ismail A, Borhan MN, Katman HY, Khalid NHN, Zaini N (2018) A case study on quality of services for bus performance in Putrajaya, Malaysia. Int J Eng Technol 7(3.9):100. Doi: https://doi.org/10.14419/ijet.v7i3.9.15825

14. Shen X, Feng S (2020) How public transport subsidy policies in China affect the average passenger load factor of a bus line. Res Transp Bus Manag 36:100526. Doi: https://doi.org/10.1016/j.rtbm.2020.100526
15. Norhisham S et al (2019) Service frequency and service hours evaluation for bus service in West Klang Valley. IOP Conf Ser Mater Sci Eng 636(1). Doi: https://doi.org/10.1088/1757-899X/636/1/012008
16. Li J, Chen X, Li X, Guo X (2013) Evaluation of public transportation operation based on data envelopment analysis. Procedia—Soc Behav Sci 96:148–155. https://doi.org/10.1016/j.sbspro.2013.08.020
17. Wang L, Li L, Wu B, Bai Y (2013) Private car switched to public transit by commuters, in Shanghai, China. Procedia—Soc Behav Sci 96:1293–1303. https://doi.org/10.1016/j.sbspro.2013.08.147
18. Kittelson I, Assoc I, Parsons Brinckerhoff, I, KFH Group, Texas A&M Transportation Institute, and Arup (2013) Transit capacity and quality of service manual, 3rd Edition [Online]. http://www.trb.org/main/blurbs/169437.aspx%5Cnhttp://www.worldtransitresearch.info/research/4941/
19. Insani TD, Handayani W, Astuti MFK, Basuki KH, Setiadji BH (2021) A performance study of bus rapid transit lite: toward a resilient Semarang City. Transp Probl 16(3):105–118. https://doi.org/10.21307/tp-2021-045
20. Ratanawaraha A, Chalermpong S (2021) Operational models, drivers' compensation, and bus service quality in Bangkok. Eng J 25(3):85–94. https://doi.org/10.4186/ej.2021.25.3.85
21. Mohamad AM, Hamin Z, Md Nor MZ, Kamaruddin S, Nizam Md Radzi MS (2020) The implications of audio/video conference systems on the administration of justice at the Malaysian Courts. Webology 17(2):904–921. Doi: https://doi.org/10.14704/WEB/V17I2/WEB17076
22. Nor MZM, Mohamad AM, Azhar A, Latif HM, Khalid AHM, Yusof Y (2019) Legal challenges of Musharakah Mutanaqisah as an alternative for property financing in Malaysia. J Leg Ethical Regul Issues 22(3).
23. Sani NM, Chun YK, Munaaim MAC (2020) Integration of UNIMAP's student bus rout ing towards free city-bus service local people in Perlis. Quantum J Soc Sci Humanit 1(5):55–68
24. Omar MF, Nawi MNM, Jamil JM, Mohamad AM, Kamaruddin S (2020) Research design of mobile based decision support for early flood warning system. Int J Interact Mob Technol 14(17):130–140. https://doi.org/10.3991/ijim.v14i17.16557

Structural Health Monitoring: A Review on Its Application in Historical Structure

Franklyn F. Manggapis and Orlean G. Dela Cruz

Abstract Historical structure is an integral component of the world's cultural identities. However, despite its cultural significance, it is the most prone type of building due to environmental factors such as aging of the materials, the effect of temperature, soil condition, and natural disasters such as earthquakes and typhoons. Therefore, the preservation of historical structure is one of the growing interests in recent years, and in monitoring the historical structure's state correctly, different methods and tools introduced. One of the methods in determining the health of the historical structure implemented a decade ago is Structural Health Monitoring (SHM). Generally, there are two types of SHM, long-term monitoring or Static SHM, which measures slow varying factors, and Dynamic SHM which determines the dynamic properties of the structure. However, with the continuous advancement of the SHM, the uncertainty and inaccuracy of the model and results are still the most significant gap in the application of SHM. This paper aims to review some of the applications of SHM in the preservation and monitoring of historical structure to provide knowledge about the topic and determine gaps and challenges based on the existing literature and studies.

Keywords Structural Health Monitoring · Wireless Sensor Networks · Structural integrity · Structural stability · Historical structure

1 Introduction

The state of a structure should be assessed similarly to that of a human being, with doctors evaluating human health using medical knowledge and advanced technology. On the other hand, engineers utilize Structural Health Monitoring (SHM) that employs modern sensors to assess the structural integrity and durability based

F. F. Manggapis
Infrastructure Development Office, Technological University of the Philippines, Taguig, Philippines
e-mail: franklyn.manggapis@tup.edu.ph

F. F. Manggapis · O. G. D. Cruz (✉)
Graduate School, Polytechnic University of the Philippines, Manila, Philippines
e-mail: ogdelacruz@pup.edu.ph

E. M. Nia et al. (eds.), *Sustainable Development Approaches*, Lecture Notes in Civil Engineering 243, https://doi.org/10.1007/978-3-030-99979-7_4

on the information obtained. In any case, this evaluation will help to identify problems early on, and engineers can provide guidelines and recommendations. Overall, the fundamental goal of structural health monitoring is to observe in-situ structural behavior under various loading circumstances during a defined period or the structure's lifetime and to detect aggressive environmental conditions [1].

One such structure most applicable to utilizing SHM systems is the Historical Structure, as these are irreplaceable assets. Furthermore, it is a valuable resource of pride and symbol of a country's cultural history worldwide. Therefore, its upkeep and preservation necessitate striking a balance between structural safety and architectural value. This paper attempts to present the viability of applying the SHM system in determining the condition of Historical Structure.

2 Research Methodology

This paper follows the work introduced by Tranfield et al. [2]. This methodology utilizes three stages, namely: (a) formulating a research question, (b) Conducting the review, and (b) Reporting the review. The basis of quality study is a good research question, which is crucial in gathering information, gaining insight into a particular problem [3], identifying the topic of interest, and a guide for methodology [4]. Therefore, researchers have developed the following research question to achieve the objective of this study:

RQ 1: What are the classifications of Structural Health Monitoring?
RQ 2: What are the instruments used in Structural Health Monitoring?
RQ 3: What are the applications of SHM in Historical Structure?
RQ 4: What are the challenges and development of SHM?

The steps in finding a paper that is related to the topics are as follows: (a) established a keyword, (b) database searching using a Boolean syntax which enables users to blend keywords such as AND, OR, and NOT using "Title/Abstract/Keyword" field of the database [5], and lastly (c) document the paper based on its eligibility for the topic by using Preferred Reporting Items for Systematic reviews and Meta-Analyses guide [6], which illustrates in Fig. 1.

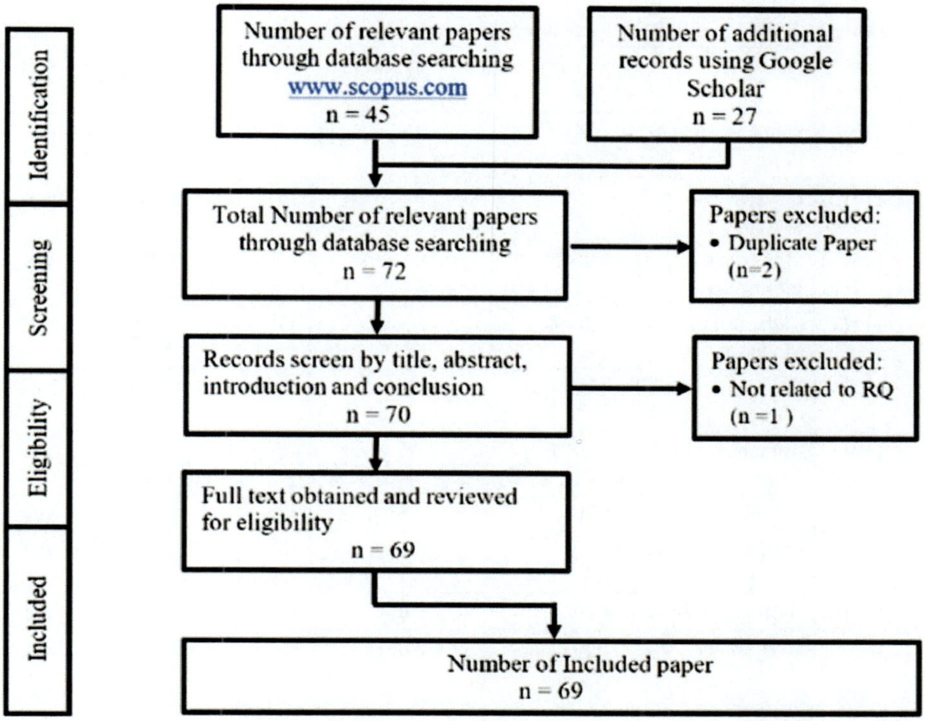

Fig. 1 Systematic related literature selecting process

3 Static and Dynamic Structural Health Monitoring

The building's structural integrity and durability require regularly monitoring especially ancient structures since it is the most vulnerable type of structure due to its deteriorating age and uncertainties in its material behavior. The traditional way of assessing the structure's health is primarily done through visual inspection by technical experts and engineers; however, this is inefficient in assessing the building's real-time condition. SHM has proven to be a powerful tool in addressing this issue.

Generally, there are two main classifications of Structural Health Monitoring: (a) Static SHM and (b) Dynamic SHM. The latter identifies that the dynamic structural reaction must be accounted for by a continuous data acquisition when measuring events like earthquake movements or traffic-related vibrations [7–9]. Conversely, the former involves continuous monitoring of critical slow-varying indicators such as inclination, corrosion, variation in time of cracks opening, settlement, humidity, and temperature [10, 11].

SSHM's fundamental goal is to determine whether the structure under observation is stable. A steady condition implies that the structure is safe; however, a non-stationary response may signal a state of deterioration, thus jeopardizing the monument's structural safety. DSHM, on the other hand, provides the dynamic properties

Table 1 SSHM and DSHM application in the Historical Structure

Ref	Structure	Year it started	Duration (years)	No. of Instrument	SSHM	DSHM
[12]	Santa Maria del Fiore	1987	20 years	150	Yes	No
[13]	Basilica of San Marco	1991	3.5 years	23	Yes	No
[14]	Monastery of Jeronimos	2005	9 years	10	Yes	Yes
[15]	Cathedral of Moderna	2004	8 years	25	Yes	No
[15]	The two towers of Bologna	2011	4 years	58	Yes	No
[16]	Roman Arena of Verona	2011	1.5 years	40	Yes	Yes
[15]	Asinelli Tower	2012	3 months	4	No	Yes
[17]	Baptisery of San Giovanni	2013	3 days	10	No	Yes
[18]	Giotto Bell Tower	2013	3 days	10	No	Yes
[8]	Sciri Tower	2017	8 days	7	No	Yes

of the structure to develop a mathematical model of the building's behavior through theoretical and experimental modal analysis [19]. Some of SSHM and DSHM in historical structure as shown in Table 1. It is noticeable that the duration of static monitoring would take years since it monitors the stationary condition of the building, and it is sometimes called continuous monitoring. Nevertheless, the structure is not limited to only one type of monitoring; in some instances, the structure uses both monitorings, such as in [14, 16].

4 Structural Health Monitoring Sensor

SHM systems contain a collection of small detachable sensors which forms to monitor either the long-term evolution of fractures, settlements, inclinations or dynamic qualities such as frequencies and damping ratios [20]. The SHM's central heart is the sensor since it is necessary for evaluating the structure; hence, the data acquisition element of the structural health monitoring process includes determining the types of sensors, and their use, The main types of sensors used in SHM are (a) Fiber Optic Sensor and (b) Wireless Sensor Network (WSN).

Fiber optic sensor is a device used to detect different parameters such as temperature [7], structural vibrations [8–10], displacement [15, 17], acceleration [21], and rotations. A fiber optic sensor system is composed of fiber optic cable linked to a

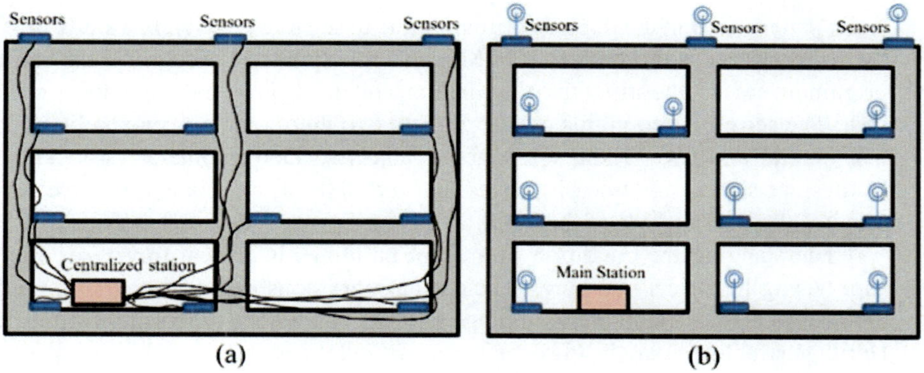

Fig. 2 SHM system set up: **a** Traditional SHM location, and **b** WSN [23]

remote sensor or an amplifier. Some examples of this sensor are (a) an accelerometer for measuring acceleration, (b) a thermometer that measures the temperature that affects the structure's physical properties, and (c) Inclinometers to monitor the subsurface movement and the slope position of the buildings.

Wireless Sensor Networks (WSN) offer the same capability as another sensor at a cheaper cost, allowing for considerably denser monitoring. The difference between utilizing typically wired sensors and wireless systems in SHM is that the latter features sensor nodes that require little maintenance and no wires, allowing them to deploy previously impractical or inaccessible areas [22].

Figure 2a shows the traditional way of installing the sensor in the building for SHM. In this approach, wired sensors may be costly and impractical to use in extensive infrastructure due to economic factors. In recent years, experts have found a way to address the problem using a wireless sensor that connects the sensor to its base station, where all the data keep in place and ready for analysis, as shown in Fig. 2b.

5 Structural Health Monitoring in Historical Structure

The historical structure has been part of human history throughout the centuries; this type of structure is prone to cataclysmic events that can compromise the stability and safety of the structure, mainly degradation, environmental hazards, and aging [24]. Therefore, increasing interest in structural health monitoring (SHM) as a knowledge-based evaluation method to measure and mitigate uncertainty about structural performance of cultural heritage sites has resulted from the need for successful seismic safety and vulnerability assessment [25].

In the study [26], five high-sensitivity accelerometers, three at the base of the drum and two at the Basilica of Saint Mary of the Angels, performed a dynamic identification to implement an effective system for early detection of damage. The affected zone discovered by numerical analysis has established that the dynamic crack pattern seen in the dome is primarily due to seismic activity. In a study in 2013, the SHM system at the Cathedral of Modena in Italy aims to determine the church's prominent cracks, the structure's inclination, and the displacement it induces over the year. The sensors used in this monitoring are two thermometers, two inclinometers, five biaxial and two Trixial joint meters, and two Deformometer [25]. All these studies are significant since it proves that the SHM system is an effective way to monitor the overall state of the building.

Understanding the condition of ancient buildings is critical for preserving and supervising historical structures; ancient masonry constructions make up much of the historical and architectural heritage and by a wide variety of complexities [26]. Heritage structures may be classified as a particular case because they are structural construction that differs from modern ones, with nuanced behavior that is often impossible to measure or understand using existing rules, guidelines, procedures, or devices. Despite the advancement and innovation of the past researcher in SHM, there are still no clear international codes for this topic [27].

6 Challenges and Gap

Considering the evolution of the SHM system in assessing the stability of not just ordinary structure but also historic structure, it is safe to state that SHM is an efficient way to monitor. However, despite the benefits, the limitations and issues must still need to address. Although the experimented investigations of the applications are successful, theoretical and practical issues still hamper a large scale of continuous monitoring. One example to be considered barriers in implementing the SHM is the number of sensors to be installed in the structure [7], this would result in the high cost of fabrication and installation of the system wherein the stakeholders, or the client may have refused the project due to the low return of investment. It is also worth noting that most of the available literature on the application of SHM in historical structure can be located in Mediterranean Europe, mainly in Italy [7–11, 19, 28, 29]; this is because most of the landmark of cultural sites are present in the said part of the world. Focusing more on the objective of this paper, Table 2 shows the challenges imposed by SHM application in historical structure based on the literature gathered.

The vibration-based method has been the most frequent technique in monitoring the structure's condition, and this method is impossible without the instrument; hence it is the most challenging part of implementing SHM. It may include insufficient and unreliable data for modeling, the location of the sensors, and the cost of the device. Similarly, the model may cause uncertainties due to material degradation, operational factors, and environmental factors. The following factors define the uncertainties in the output and result of the SHM: (a) measurement errors, (b) site conditions, (c) calibration and tolerance, (d) transmission and storage issue, and (e) final monitoring are inconclusive.

Table 2 Challenges and Limitations of SHM in Historical Structure

References	Challenges and Limitations
[7, 8, 28, 30–33]	The uncertainty in the output of the instrument
[34, 35]	The uncertainty in the result of model
[9]	Difficulties in where to install the sensors
[36]	Synchronization of wireless sensor network
[24]	The lack of international code and guidelines for SHM
[9]	Disapproval of the usage of SHM due to financial reason

7 Conclusion

Preservation and maintenance of an ancient structure is a daunting task that engineers and experts need to address. SHM is an excellent tool in monitoring the structure's health, whether short-term (DSHM) or long-term (SSHM); the output of this method significantly affects the engineer's decision for preventive measures. In finding the result of the monitoring, sensors with a centralized panel shall be installed on the site.

The duration of the monitoring depends on the type of monitoring, and this will take days, months, years, or continuous depending on what method requires to utilize. Although this monitoring method is beneficial to the existing structure, several gaps and challenges still exist, such as the uncertainties in the instrument's output due to the wrong placement of sensors or not-in-sync device; because of this, the model that forms in the result is questionable. However, the main problem in utilizing the SHM comes with financial reasons; moreover, this is applicable most of the time in developed countries. To conclude this paper, the development of SHM is significant in historical structure since this will be an excellent strategy in minimizing catastrophic losses by foreseeing its damage in the early stages.

References

1. Farrar CR, Worden K (2010) An introduction to structural health monitoring. Doi: https://doi.org/10.1007/978-3-7091-0399-9_1
2. Tranfield D, Denyer D, Smart P (2003, September) Towards a methodology for developing evidence-informed management knowledge by means of systematic review. Br J Manag 14(3). Doi: https://doi.org/10.1111/1467-8551.00375
3. Kishore J, Vasundhra S, Anand T (2011, July) Formulation of a research question. Indian J Med SpecTies 2(2):184–188
4. Ratan S, Anand T, Ratan J (2019) Formulation of research question—Stepwise approach. J Indian Assoc Pediatr Surg 24(1). Doi: https://doi.org/10.4103/jiaps.JIAPS_76_18

5. dela Cruz OG, Mendoza CA, Lopez KD (2021, July) International Roughness Index as Road Performance Indicator: A Literature Review. IOP Conf Ser: Earth Environ Sci 822(1):012016. Doi: https://doi.org/10.1088/1755-1315/822/1/012016
6. Moher D, Liberati A, Tetzlaff J, Altman DG (2009, July) Preferred reporting items for systematic reviews and meta-analyses: the PRISMA statement. PLoS Med 6(7). Doi: https://doi.org/10.1371/journal.pmed.1000097
7. Simonetta B, Michele P, Silvestri S, Gasparini G, Trombetti T (2015, July) SSHM and DSHM for a better knowledge and risk prevention of historical buildings: the cases of the Two Towers in Bologna and the Cathedral in Modena. Doi: https://doi.org/10.1109/EESMS.2015.7175874
8. Kita A, Cavalagli N, Comanducci G, Ubertini F (2017) Dynamic testing and monitoring of historic towers for seismic damage detection. Proceedings 6th Int Conf Comput Methods Struct Dyn Earthq Eng 1:2564–2577. Doi: https://doi.org/10.7712/120117.5589.18130
9. Ceravolo R, Pistone G, Fragonara LZ, Massetto S, Abbiati G (2016) Vibration-based monitoring and diagnosis of cultural heritage: a methodological discussion in three examples. Int J Arch Herit 10(4):375–395. https://doi.org/10.1080/15583058.2013.850554
10. Baraccani S, Palermo M, Azzara RM, Gasparini G, Silvestri S, Trombetti T (2017, July) Structural interpretation of data from static and dynamic structural health monitoring of monumental buildings. Key Eng Mater, vol 747. Doi: https://doi.org/10.4028/www.scientific.net/KEM.747.431
11. Makoond N, Pelà L, Molins C, Roca P, Alarcón D (2020, October) Automated data analysis for static structural health monitoring of masonry heritage structures. Struct Control Health Monit 27(10). Doi: https://doi.org/10.1002/stc.2581
12. Ottoni F, Blasi C (2015, January) Results of a 60-year monitoring system for Santa Maria del Fiore Dome in Florence. Int J Arch Herit 9(1). Doi: https://doi.org/10.1080/15583058.2013.815291
13. Rossi PP, Rossi C (2015, January) Monitoring of two great venetian cathedrals: San Marco and Santa Maria Gloriosa Dei Frari. Int J Arch Herit 9(1). Doi: https://doi.org/10.1080/15583058.2013.793435
14. de Stefano A, Matta E, Clemente P (2016, February) Structural health monitoring of historical heritage in Italy: some relevant experiences. J Civ Struct Health Monit 6(1). Doi: https://doi.org/10.1007/s13349-016-0154-y
15. Baracanni S, Trombetti T, Palermo M, Gasparini G, Silvestri S, Dib A (2014, July) A methodology of analysis for a critique interpretation of the data acquired from monitoring systems of historical buildings. 7th Eur Work Struct Health Monit, pp 655–662
16. Masciotta M-G, Roque JCA, Ramos LF, Lourenço PB (2016, July) A multidisciplinary approach to assess the health state of heritage structures: the case study of the Church of Monastery of Jerónimos in Lisbon. Constr Build Mater, vol 116. Doi: https://doi.org/10.1016/j.conbuildmat.2016.04.146
17. Lacanna G, Ripepe M, Marchetti E, Coli M, Garzonio CA (2016, July) Dynamic response of the Baptistery of San Giovanni in Florence, Italy, based on ambient vibration test. J Cult Herit, vol 20. Doi: https://doi.org/10.1016/j.culher.2016.02.007
18. Lacanna G, Lancellotta R, Ripepe M (2019) Integrating modal analysis and seismic interferometry for structural dynamic response the case study of giotto's bell tower in Florence (Italy). Doi: https://doi.org/10.7712/120119.7036.18799
19. Tronci EM, de Angelis M, Betti R, Altomare V (2020, December) Vibration-based structural health monitoring of a RC-masonry tower equipped with non-conventional TMD. Eng Struct, vol 224. Doi: https://doi.org/10.1016/j.engstruct.2020.111212
20. Sohn H, Farrar C, Hemez F, Czarneck J (2002, December) A review of structural health monitoring literature 1996–2001. Los Alamos Natl Lab (LANL), pp 1–7
21. Lorenzoni F, Caldon M, da Porto F, Modena C, Aoki T (2018, April) Post-earthquake controls and damage detection through structural health monitoring: applications in l'Aquila. J Civ Struct Health Monit 8(2). Doi: https://doi.org/10.1007/s13349-018-0270-y
22. Avci O, Abdeljaber O, Kiranyaz S, Hussein M, Inman DJ (2018, June) Wireless and real-time structural damage detection: a novel decentralized method for wireless sensor networks. J Sound Vib, vol 424. Doi: https://doi.org/10.1016/j.jsv.2018.03.008

23. Spencer BJr, Ruiz-Sandoval M, Kurata N (2004) Smart sensing technology for structural health monitoring
24. Mesquita E, Antunes P, Coelho F, André P, Arêde A, Varum H (2016, July) Global overview on advances in structural health monitoring platforms. J Civ Struct Health Monit 6(3). Doi: https://doi.org/10.1007/s13349-016-0184-5
25. Lorenzoni F, Casarin F, Caldon M, Islami K, Modena C (2016, January) Uncertainty quantification in structural health monitoring: applications on cultural heritage buildings. Mech Syst Signal Process, vol 66–67. Doi: https://doi.org/10.1016/j.ymssp.2015.04.032
26. Cavalagli N, Botticelli L, Gioffrè M, Gusella V, Ubertini F (2017) Dynamic monitoring and nonlinear analysis of the dome of the basilica of S. Maria degli Angeli in Assisi. Doi: https://doi.org/10.7712/120117.5587.18117
27. de Stefano A (2007, September) Structural identification and health monitoring on the historical architectural heritage. Key Eng Mater , vol 347. Doi: https://doi.org/10.4028/www.scientific.net/KEM.347.37
28. Makhoul N (2018) Preservation of an existing original building by studying its dynamic properties. Doi: https://doi.org/10.2749/nantes.2018.s6-1
29. Zonno G, Aguilar R, Boroschek R, Lourenço PB (2018, November) Automated long-term dynamic monitoring using hierarchical clustering and adaptive modal tracking: validation and applications. J Civ Struct Health Monit 8(5). Doi: https://doi.org/10.1007/s13349-018-0306-3
30. Antunes P et al (2011) Optical sensors based on fiber bragg gratings for structural health monitoring. Doi: https://doi.org/10.1007/978-3-642-21099-0_12
31. Barrias A, Rodriguez G, Casas JR, Villalba S (2018, July) Application of distributed optical fiber sensors for the health monitoring of two real structures in Barcelona. Struct Infrastruct Eng 14(7). Doi: https://doi.org/10.1080/15732479.2018.1438479
32. Wang J, Chen H, Du X (2020, December) Study on the early warning mechanism for real-time monitored structural responses of a historical timber building. Measurement 165. Doi: https://doi.org/10.1016/j.measurement.2020.108136.
33. Bacco M et al (2020) Monitoring ancient buildings: real deployment of an IoT system enhanced by UAVs and virtual reality. IEEE Access, vol 8. Doi: https://doi.org/10.1109/ACCESS.2020.2980359
34. Burgos M, Castaneda B, Aguilar R (2019) Virtual reality for the enhancement of structural health monitoring experiences in historical constructions. Doi: https://doi.org/10.1007/978-3-319-99441-3_46
35. Pachón P et al (2020, January) Evaluation of optimal sensor placement algorithms for the Structural Health Monitoring of architectural heritage. Application to the Monastery of San Jerónimo de Buenavista (Seville, Spain). Eng Struct, vol 202. Doi: https://doi.org/10.1016/j.engstruct.2019.109843
36. Bezas K, Komianos V, Oikonomou K, Koufoudakis G, Tsoumanis G (2019, September) Structural health monitoring in historical buildings using a low cost wireless sensor network. Doi: https://doi.org/10.1109/SEEDA-CECNSM.2019.8908531

Assessment of Customers' Satisfaction in Mass Rapid Transit (MRT) Using Importance-Performance Analysis (IPA)

Nur Sarah Shaziah Samsudin, Norlela Ismail, Shuhairy Norhisham, Nur'atiah Zaini, Muhammad Fadhlullah Abu Bakar, Nor Najwa Irina Mohd Azlan, and Ron Aldrino Chan alias Ron Buking

Abstract In developing countries, public transit use is still lacking as opposed to developed counties. Therefore, it is essential to consider consumers' public transportation service quality expectations to increase overall customer experience and Mass Rapid Transit (MRT) ridership in Malaysia. Thus, this article's objective is to suggest thresholds for assessing customers' satisfaction of service quality in Malaysia's MRT. The questionnaire collected responses from MRT users. For methodology, the dimensions considered in this analysis were primarily based on the five (5) SERVQUAL dimensions (Reliability, Assurance, Tangibility, Empathy, and Responsiveness) with the addition of Safety & Security and Accessibility dimension. As for the results, the reliability test shows that Tangibility has the highest Cronbach value in performance with 0.881, while Responsiveness is the highest in importance with a Cronbach alpha value of 0.971. Importance—Performance Analysis (IPA) to indicate customers' satisfaction level of service quality was included. The mean importance (4.28) scored higher than the mean performance (4.00). This article concludes that the IPA results show that the responsiveness dimension should be prioritised for improvement. Furthermore, this research could contribute to MRT operators, researchers, and policymakers to formulate strategies that will increase MRT use in Malaysia.

Keywords Mass Rapid Transit · Quality of service · Rail transportation · Importance-Performance Analysis (IPA) · Public transportation

---

N. S. S. Samsudin (✉) · N. Ismail · S. Norhisham · N. Zaini · M. F. A. Bakar · N. N. I. M. Azlan
Department of Civil Engineering, College of Engineering, Universiti Tenaga Nasional, 43000 Kajang, Selangor, Malaysia
e-mail: sarahshaziah21@gmail.com

N. Ismail · S. Norhisham · N. Zaini
Institute of Energy Infrastructures (IEI), Universiti Tenaga Nasional, 43000 Kajang, Selangor, Malaysia

R. A. C. Ron Buking
Faculty of Engineering, Department of Civil Engineering, Universiti Malaysia Sarawak, 94300 Kota Samarahan, Sarawak, Malaysia

E. M. Nia et al. (eds.), *Sustainable Development Approaches*, Lecture Notes in Civil Engineering 243, https://doi.org/10.1007/978-3-030-99979-7_5

1 Introduction

The urban rail transit system has become one of the most preferred modes of transportation when getting around in the city centre [1]. It seems to be the best way to reduce travel time, especially around central business and commercial districts where congestion is usually the problem. Many developed cities, such as London, Paris, Seoul, and China, have successfully controlled their traffic flow to minimise road travel time for motorists [2]. In Malaysia, public transportation is currently growing, especially rail transportation in connecting the urban areas to the city centre where the central business and commercial districts are located [3, 4].

Mass Rapid Transit (MRT) is one of the most convenient transportation modes for urban commuters and it offers the latest technology as compared to other types of transit. The first MRT line was introduced as the Sungai Buloh - Kajang Line (currently known as Kajang Line) which has a total distance of 46 km. This fast-paced development is due to the Government's initiative as it can reduce traffic congestion, air pollution, and the cost of daily commute [5–7]. However, in Malaysia, the public seems less keen to use it than developed countries due to their experience with public transportation's performance and reliability [8, 9]. The facilities and quality of services provided could be factors causing commuters to prefer other modes of transportation such as private vehicles [10, 11].

The stated reasons above may have significantly affected users' confidence in using the MRT and may have even led them to avoid taking it altogether, which might explain the lower than projected ridership figure. As a result, the Government's objectives might take longer or even would not materialise if no timely action is taken to address these issues [12]. While the service quality of public transportation in Malaysia, such as bus, LRT, KTM, and Monorail, has been well documented [5, 13] the customers' perception on MRT service quality is rarely discussed. Hence, this study aims to provide indication to stakeholders for improving MRT services to meet customers' expectations.

This article is organised into five sections. Section 2 discusses the sampling techniques, sample size and questionnaire design. Section 3 reveals the three methodologies used in this study. In Sect. 4, the outcome of the analysis is presented in a way to facilitate future investigations of the subject. Lastly, Sect. 5 outlines the conclusions of the study and suggests improvement for MRT operators.

2 Sampling Techniques, Sample Size and Questionnaire Development

In this study, the convenience sampling of non-probability sampling technique is used. It is widely used among students as it is inexpensive, less complicated, and easier to implement compared to probability sampling [14]. A pilot study was

Table 1 Summary of adapted attributes in its modified SERVQUAL dimension [17]

Dimension	Number of items	Attributes
Reliability	5	G1 MRT always arrives on time/punctual G2 MRT rarely breaks down/stop on the track G3 Staff satisfies customers' request right the first time G4 MRT schedule timetable is easily available/accessible G5 MRT status are frequently updated
Assurance	5	G6 Customers' belongings are secure G7 Staff have in-depth occupational knowledge of their jobs G8 Behaviour of staff in-stills confidence in the customers G9 MRT station and coach are comfortable and not crowded G10 entilation and air-conditioning system work well
Tangibles	5	G11 MRT staffs have a professional appearance G12 The ticket office is attractive and neat G13 MRT station are well maintained and clean G14 MRT coach is always clean G15 Other facilities and amenities are in decent condition (toilet, prayer room, etc.)
Empathy	5	G16 Willingness of staff to address customers' queries G17 Suitable operating hours for all customers G18 Easy accessibility of information about services G19 Easy to find and access the ticket office/machine G20 Customers can book/purchase tickets easily
Responsiveness	4	G21 MRT company often notify customers in advance about the provision of services and price adjustments G22 Availability of staff to assist customer G23 Complaint is handled promptly and in an orderly manner G24 Precise and accurate answer to customers' queries
Safety and security	3	G25 There are guards/auxiliary police around the station during operating hours G26 Presence of CCTV G27 Stations are well-lit especially in the evening/night
Accessibility	5	G28 There are lifts to the station platform G29 Wheelchairs are available for elderly G30 Have adequate facilities for the disabled G31 Park N Ride facility is available G32 Excellent feeder services connecting to other public transport

conducted with the selection of 30 respondents as adapted from a similar study [15].

A sample is a group of a smaller number of persons drawn from a larger population for study. According to A. Mat et al., there are no specific recommendations for selecting sample sizes [16]. As a result, anecdotal evidence revealed that each questionnaire requires at least 100 to 200 respondents.

The questionnaire contains two main segments. Respondents were required to score the attributes on a five-point Likert scale, from 1 = strongly disagree to 5 = strongly agree. Table 1 shows the summary of attributes inclusive of each dimension with the use of 'G' as an indicator in this study.

3 Research Methodology

When developing research questions, the relevance of the questionnaire has to be assessed by researchers [18]. Cronbach's alpha (a) is often used to analyse the reliability of the questionnaires developed [19]. The result will be in a range from 0 to 1, with 1 being the most reliable. If the value is less than 0.6, the questions for the dimension is classified as unsatisfactory and the dimension will not be analysed as it does not affect the outcome [20].

Raw data is transformed into usable data such as frequency distribution, percentage, mean, and standard deviation. This will leave the data organised and ready to be analysed further. In a recent study on the elderly users' perception of public transportation in Thailand, the socio-demographic and travel behaviour were collected as descriptive analysis [21]. These authors used percentages and frequencies to represent the characteristics of respondents in their results [22, 23]. However, in addition to the respondents' socio-demographic data collected, these researchers described and relate it with the perceived quality services results.

The mean and statistical analysis was performed to assess the gap between passengers' needs and satisfaction with MRT service quality in the second segment of the questionnaire, where the five-point Likert scale was implemented. The scale has four quadrants and each quadrant has their own element which is based from the interest-performance calculation data. The Importance-Performance Analysis (IPA) or quadrant analysis developed by Martilla and James published in 1977 [24] is vastly used in understanding customers' satisfaction [8].

Table 2 Reliability test results of each dimension

Dimension	Performance	Importance
	Cronbach's Alpha	Cronbach's Alpha
Reliability	0.777	0.904
Assurance	0.801	0.882
Tangibility	0.881	0.901
Empathy	0.863	0.899
Responsiveness	0.856	0.917
Safety & Security	0.727	0.845
Accessibility	0.819	0.903

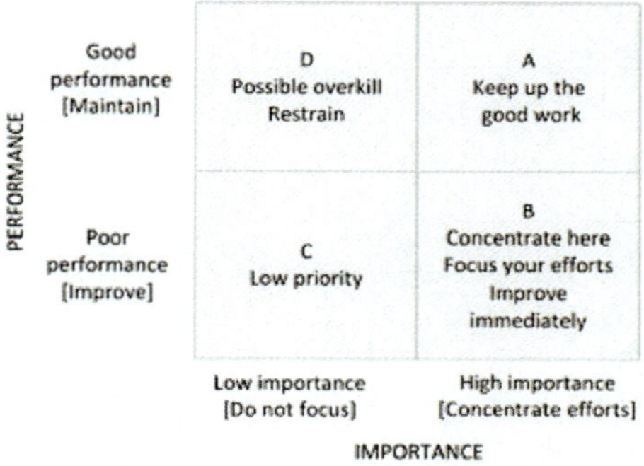

4 Results and Discussion

In this section, the results were presented with the use of analyses including Reliability test, Descriptive analysis, and Importance-Performance Analysis (IPA).

4.1 Reliability Test

Table 2 shows Cronbach's alpha values after analysing respondents' views regarding this study. The total number of respondents analysed is 137. By observing the table, it indicates that the values obtained are relatively high and reliable.

4.2 Demographic Analysis

Table 3 shows the demographic analysis of respondents in terms of frequency and percentage that were willing to participate in this questionnaire.

Table 3 Demographic analysis of respondents

Question	Responses	Frequency (n)	Percentage (%)
Age range	- Below 20	2	1.5
	- 21–35	53	38.7
	- 36–50	22	16.1
	- 51–60	38	27.7
	- Above 60	22	16.1
Gender specification	- Male	56	40.9
	- Female	81	59.1
Education level	- Less than secondary school graduate	0	0
	- SPM cert receiver or equivalent	15	10.9
	- Graduate work or degree	10	7.3
	- Diploma	12	8.8
	- Bachelor degree	71	51.8
	- Post-Graduate	29	21.2
Occupational status	- Student	36	26.3
	- Employed	59	43.1
	- Unemployed	8	5.8
	- Retired	34	24.8
	- Disabled, not able to work	0	0
Monthly Income	- Less than 1,000	47	34.4
	- 1,001–5,000	43	31.4
	- 5,001–10,000	27	19.7
	- 10,001 and more	20	14.6

4.3 Importance-Performance Analysis (IPA)

Table 4 summarises each attribute's mean performance and mean importance score obtained from respondents who contributed to the questionnaire. Methods that were mentioned in Sect. 3 were implemented in each IPA figure.

Figure 1 shows that, on average, five out of seven dimensions lie in quadrant A "Keep up the good work", which implies that most of the dimensions are at a very satisfactory level. The remaining dimensions, responsiveness, and assurance are classified in quadrant C, "Low priority" for the former, and quadrant D, "Possible overkill" for the latter. Hence, this indicates that customers are dissatisfied with the responsiveness dimension features but do not believe they are significant. As for assurance, the results show that customers were pleased with this dimension and did not perceive it to be a severe problem.

Table 4 Average performance and importance outcome of MRT services

Attributes	Mean performance	Mean importance	Gap
Reliability	4.12	4.29	−0.17
Assurance	3.93	4.22	−0.29
Tangibles	4.09	4.31	−0.22
Empathy	4.15	4.33	−0.18
Responsiveness	3.71	4.17	−0.46
Safety & Security	4.05	4.37	−0.32
Accessibility	3.94	4.30	−0.39
Total	4.00	4.28	–

Note 'G' is used as an indicator in IPA charts

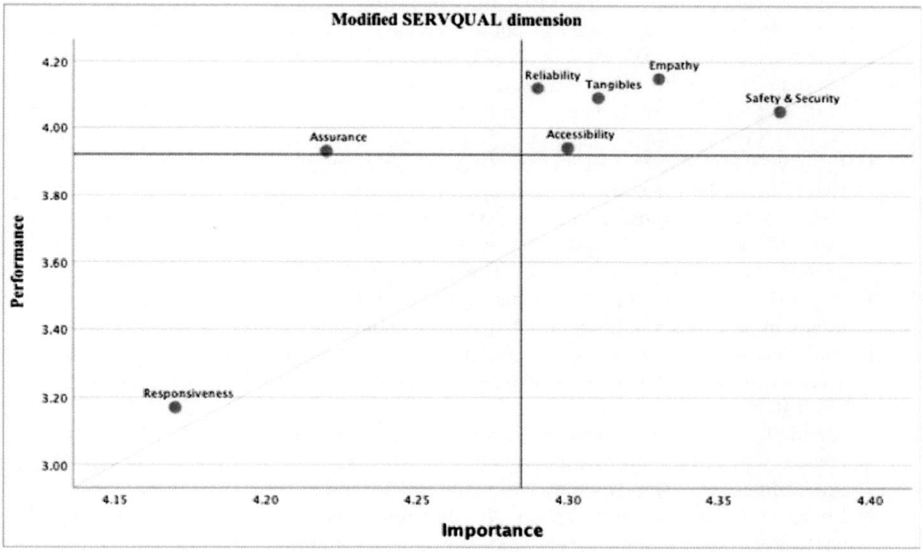

Fig. 1 IPA chart of overall modified SERVQUAL dimensions

5 Conclusion

This study sought to provide stakeholders recommendations for improving MRT services to meet customers' expectations. It reveals the overall service quality based on 32 attributes divided into seven (7) dimensions collected from various literature reviews and altered appropriately. The Cronbach alpha value obtained shows that all dimensions were reliable and analysed accordingly.

The mean importance and mean performance of each dimension were tabulated from the questionnaire results and plotted in the IPA to determine customers' satisfaction. It was discovered that the overall service quality offered by MRT is satisfactory but there is still room for improvement. To enhance its service quality, MRT could redeployed some of its resources in the assurance dimension to the responsiveness dimension since these dimensions were deemed to be 'possible overkill' and 'low priority' respectively. This will not only serve existing riders' interests, but it will also attract new users, resulting in greater use of mass rapid transit transportation.

Acknowledgements The first author would like to acknowledge UNITEN BOLD 2021 Grant J510050002/2021167 for funding the opportunity to publish this article.

References

1. Nordin NHB, Mohd Masirin MIH, Bin Ghazali MI, Bin Azis MI (2017) Appraisal on rail transit development: a review on train services and safety. IOP Conf Ser Mater Sci Eng 226(1). Doi: https://doi.org/10.1088/1757-899X/226/1/012034
2. Nordin NH, Mohd Masirin MI, Ghazali MI, Azis MI (2016) Passenger rail service comfortability in Kuala Lumpur urban transit system. MATEC Web Conf, vol 47. Doi: https://doi.org/10.1051/matecconf/20164703011
3. Norhisham S et al (2021) Evaluating the quality of services for bus performance in Alor Setars. IOP Conf Ser Earth Environ Sci 708(1). Doi: https://doi.org/10.1088/1755-1315/708/1/012038
4. Tu F, Ho CS, Teknologi U, Xiaofend M, Professor Y Promoting urban sustainability through green technology in Malaysia sustainable cities program, Working Paper series promoting urban sustainability through green technology in Malaysia
5. Das AM, Ladin MA, Ismail A, Rahmat ROK (2013) Consumer's satisfaction of public transport monorail user in Kuala Lumpur. J Eng Sci Technol 8(3):272–283
6. Norhisham S et al (2019) Service frequency and service hours evaluation for bus service in West Klang Valley. IOP Conf Ser Mater Sci Eng 636(1):7. https://doi.org/10.1088/1757-899X/636/1/012008
7. Ramli MZ et al (2017, September) Level of service for pedestrian movement towards the performance of passenger information in public transport stations in Klang Valley. AIP Conference Proceedings, vol 1885. Doi: https://doi.org/10.1063/1.5002305
8. Ibrahim, Borhan, Zakaria, Zainal (2019) Effectiveness of commuter rail service toward passenger's satisfaction: a case study from Kuala Lumpur, Malaysia. Int J Eng Technol 8(1.2):50–55. Doi: https://doi.org/10.14419/ijet.v8i1.2.24871
9. Norhisham S et al (2019) Assessing stakeholder's view for quality bus services in Kuala Lumpur. Int J Adv Sci Technol 28(10):359–369
10. Chiu Chuen O, Karim MR, Yusoff S (2014) Mode choice between private and public transport in Klang Valley, Malaysia. Sci World J, vol 2014. Doi: https://doi.org/10.1155/2014/394587
11. Ramli MZ et al (2017, September) Level of service for pedestrian movement towards the performance of passenger information in public transport stations in Klang Valley. AIP Conf Proc 1885(1). https://doi.org/10.1063/1.5002305
12. Almselati ASI, Rahmat RAOK, Jaafar O (2011) An overview of urban transport in Malaysia. Soc Sci 6(1):24–33. https://doi.org/10.3923/sscience.2011.24.33
13. Ghazali FNC, Ramli MZ, Abidin MZZ (2017) Assessment of passenger information system in light rail transit station in Klang Valley. Int J Sci Res Sci Eng Technol 3(5):461–465
14. Norhisham S, Ismail A, Borhan MN, Katman HY, Khalid NHN, Zaini N (2018) A case study on quality of services for bus performance in Putrajaya, Malaysia. Int J Eng Technol 7(3.9):100. Doi: https://doi.org/10.14419/ijet.v7i3.9.15825
15. Showkat N Parveen H (2017, August) Nonprobability and probability sampling. Commun Res, pp 0–9. Accessed: June 18, 2021

16. Mat A et al (2018) The influence of public commuters satisfaction towards Mass Rapid Transportation (MRT) services among Kuala Lumpur riders: a PLS-SEM approach. (IJEKM) Int J Educ Knowl Manag Fac Bus Manag 1(1):1–11
17. Thanaraju P, Khan PAM, Sivanathan S, Juhari NH (2019) Passengers' satisfaction towards railway facilities (RAILQUAL in the central region. Int J Recent Technol Eng 8(2):561–571 (Special Issue)
18. Taber KS (2018, December) The use of Cronbach's alpha when developing and reporting research instruments in science education. Res Sci Educ 48(6):1273–1296. https://doi.org/10.1007/s11165-016-9602-2
19. Norhisham S et al (2019) Statistical overview on quality bus services in Klang Valley. Int J Adv Sci Technol 28(10):370–380
20. Sathiaseelan P, Dava ADB, Mahmud B (2020) A study on determinants that effect Klang Valley (Malaysia) consumer preference in rail industry. IOP Conf Ser Mater Sci Eng 780(7). Doi: https://doi.org/10.1088/1757-899X/780/7/072014
21. Chaisomboon M, Jomnonkwao S, Ratanavaraha V (2020, November) Elderly users' satisfaction with public transport in thailand using different importance performance analysis approaches. Sustain 12(21):1–20. https://doi.org/10.3390/su12219066
22. Ibrahim ANH, Borhan MN, Yusoff NIM, Ismail A (2020, May) Rail-based public transport service quality and user satisfaction—a literature review. Promet—Traffic & Transp 32(3):423–435. Doi: https://doi.org/10.7307/PTT.V32I3.3270
23. Ibrahim ANH et al (2021, January) Gender and age do matter: exploring the effect of passengers' gender and age on the perception of light rail transit service quality in Kuala Lumpur, Malaysia. Sustain 13(2):1–18. https://doi.org/10.3390/su13020990
24. Jomnonkwao S, Champahom T, Ratanavaraha V (2020) Methodologies for determining the service quality of the intercity rail service based on users' perceptions and expectations in Thailand. Sustain 12(10). Doi: https://doi.org/10.3390/su12104259

Evaluating Mass Rapid Transit (MRT) Service Quality According to Customers' Age Group of Varying Travel Pattern

Shuhairy Norhisham, Nur Sarah Shaziah binti Samsudin, Norlela Ismail, Nurul Hani Mardi, Muhammad Fadhlullah Abu Bakar, Nor Najwa Irina Mohd Azlan, and Ron Aldrino Chan alias Ron Buking

Abstract In contrast to developed countries, developing countries still have low use of public transportation. Thus, to improve overall customer experience and MRT ridership, it is critical to address consumers' public transportation service quality expectations. This objective of this paper is to describe demographic characteristics that may affect customer perceived service quality and any variations in customer perception between different age groups. This survey collected questionnaire results from MRT users of the SBK Line. The methodology of this paper includes using the SERVQUAL dimensions (Reliability, Assurance, Tangibility, Empathy, and Responsiveness) with the inclusion of Safety & Security and Accessibility dimension. Cluster Analysis (CA) was included in this study to segregate respondents' age groups. The outcome of this analysis shows that most respondents were in the age range of 25–31, which consists of 38.7%. Overall, it is concluded that mean importance scored higher than mean performance for all clusters. The Importance-Performance Analysis (IPA) results show that the perceived service quality varies with age. However, respondents of all ages agree that the safety and security dimension is considered satisfactory. It is intended that this article will encourage stakeholders to improve the quality of MRT, especially in Malaysia, so that the country can compete on a global scale in terms of public transportation.

Keywords Mass rapid transit · Public transportation · Rail transportation · Service quality · Cluster analysis · Importance-performance analysis (IPA) · Age

S. Norhisham (✉) · N. S. S. binti Samsudin · N. Ismail · N. H. Mardi · M. F. Abu Bakar · N. N. I. M. Azlan
Department of Civil Engineering, College of Engineering, Universiti Tenaga Nasional, 43000 Kajang, Selangor, Malaysia
e-mail: shuhairy@uniten.edu.my

Institute of Energy Infrastructures (IEI), Universiti Tenaga Nasional, 43000 Kajang, Selangor, Malaysia

R. A. C. Ron Buking
Faculty of Engineering, Department of Civil Engineering, Universiti Malaysia Sarawak, 94300 Kota Samarahans, Sarawak, Malaysia

E. M. Nia et al. (eds.), *Sustainable Development Approaches*, Lecture Notes in Civil Engineering 243, https://doi.org/10.1007/978-3-030-99979-7_6

1 Introduction

As population development is growing higher due to the increasing economic demands for more human resources, the criteria of an effective and productive mass transportation system are becoming more critical [1, 2]. People from rural to urban areas have increased the need for alternate transport infrastructure because of the availability of jobs in cities. In Malaysia, public transportation is expanding, particularly rail transit, which connects metropolitan areas to the city centre, which houses the country's key business and commercial sectors.

For example, the Mass Rapid Transit (MRT) construction in Klang Valley is rapidly growing. It offers the latest technology as compared to other types of transit trains such as Light Rapid Transit (LRT) and KL Monorail [3]. The first line was introduced back in 2016 and it starts from Sungai Buloh and ends in Semantan (SBK Line). This rapid progress has the potential to have a positive impact in the transportation sector, where it has been shown to minimise the risk of traffic accidents, pollution, and global warming [4, 5].

Even though the public is generally aware of the positive impact in MRT usage, the current ridership is low with just over 25 percent of the projected ridership of 400,000 passengers daily [6]. There could be many reasons due to this shortcoming, one being the perception that the services provided by public transportation authorities are insufficient [7, 8]. There have been many instances where the existing facilities are not as reliable as expected [9], frequent reports on train delays [10], and an increase in the number of crime incidents are reported in MRT stations [11].

While the service quality of public transportation in Malaysia, such as bus, LRT, and Monorail, has been thoroughly discussed, the influence of age on MRT service quality is rarely mentioned [1]. Hence, this study aims to demonstrate the impact of different age populations on the service quality of MRT in Malaysia. It is intended that this article will encourage stakeholders to improve the quality of MRT, so that the country can compete on a global scale.

2 Data Collection

The survey is developed by using past literature reviews from similar research on customers' satisfaction [12–14], and it contains three main sections. The first section is about the respondents' general data and travel behaviour, while the second section shows the data on customers' travel patterns. The third segment is the measurement of performance and the importance of the service quality attributes to customers according to the modified SERVQUAL dimensions. Respondents were asked to rate the attributes on a five-point Likert scale, with $1 =$ strong disagreement and $5 =$ strong agreement.

The necessity of selecting an optimal sample size demonstrates the relevance of choosing a sufficient sample size for minimising the cost of sampling errors. There

Fig. 1 Traditional IPA map [20]

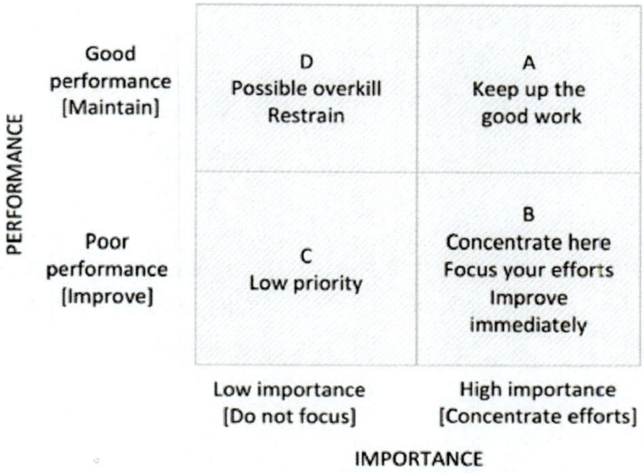

are no clear recommendations for selecting sample sizes, according to Azmi Mat [15]. Based on anecdotal data, each survey required at least 100 to 200 respondents.

3 Methods of the Research

The systematic procedure for inspecting the data includes an analysis of the respondents' descriptive analysis and Cluster Analysis (CA) as it is recommended to use CA when dealing with nonnormal variables because respondents' demographic backgrounds and experiences can affect their assessments of performance and importance, and different demographic segments can produce substantially different results [16]. The approach taken in this article is the use of CA to group survey participants who share similar age range to observe their views towards services. Cronbach's alpha (a) was adopted for the study to perform a reliability test. The outcome is a number between 0 and 1, with 1 being the most reliable [17, 18]. If the value is less than 0.6, the dimension will not be evaluated [19]. Finally, the gap between passengers' needs and satisfaction with MRT service quality in the third segment of the questionnaire was analysed, and the values were plotted into Importance-Performance Analysis (IPA) charts [20] (Fig. 1).

4 Results and Discussion

In this section, the results were presented with the use of analyses including Reliability test, Descriptive analysis, and Importance-Performance Analysis (IPA).

Table 1 Reliability test results of each dimension

Dimension	Performance	Importance
	Cronbach's Alpha	Cronbach's Alpha
Reliability	0.777	0.904
Assurance	0.801	0.882
Tangibility	0.881	0.901
Empathy	0.863	0.899
Responsiveness	0.856	0.917
Safety & Security	0.727	0.845
Accessibility	0.819	0.903
Total	0.950	0.977

4.1 Reliability Test

Table 1 shows Cronbach's alpha values after analysing respondents' views regarding this study. The total number of respondents analysed is 137, after removing eight (8) respondents who have never used MRT's services. By observing the table, it indicates that the values obtained are relatively high and reliable.

4.2 Descriptive Analysis

In this part of the study, the distribution of results was analysed according to respondents' age group.

4.2.1 Section 1—Demographic analysis

Table 2 shows a summary of the demographic analysis of every cluster. These options were chosen to be analysed and further relate with their views on provided quality services in MRT using the IPA matrix.

4.2.2 Section 2—Travel Pattern

In Cluster 2, most of the respondents use MRT to Shop or Eating out (42%), followed by To or from work (36%). To or from work ranks first in trip purpose for Cluster 3 with 36%. Recreation/Social/Place of worship and Shopping or Eating outranks high up in Cluster 3 with 32% and 23% respectively. The remainder of the respondents in Cluster 3 chose others for their commute purpose.

Recreation/Social/Place of worship, shopping/eating out, and others rank high in Cluster 4 with the same percentage of 12%, while only 4% uses MRT as their work

Table 2 Demographic analysis of respondents

Question	Responses	Below 20 (C1)		21–35 (C2)		36–50 (C3)		51–60 (C4)		Above 50 (C5)	
		N	%	N	%	N	%	N	%	N	%
Gender specification	– Male	0	0	11	21	12	55	18	47	15	68
	– Female	2	100	42	79	10	45	20	53	7	32
Occupational status	– Student	2	100	34	64	0	0	0	0	0	0
	– Employed	0	0	19	36	18	82	20	53	2	9
	– Unemployed	0	0	0	0	2	9	3	8	3	14
	– Retired	0	0	0	0	2	9	15	39	17	77
	– Disabled	0	0	0	0	0	0	0	0	0	0
Monthly income	– Less than 1,000	1	50	34	64	1	5	7	18	4	18
	– 1,001–5,000	1	50	15	28	8	36	10	26	9	41
	– 5,001–10,000	0	0	4	8	6	27	12	32	5	23
	– 10,001 and more	0	0	0	0	7	32	9	24	4	18

* N = Frequency, % = Percentage

commute. In the last cluster (C5), where most of these respondents were retired, they mostly use the MRT to connect to their Recreation/Social/Place of worship location (32%). While analysing these figures, it can be concluded that most of the respondents use MRT to Shop/Eat out.

Next, 45% of Cluster 2 respondents chose MRT to travel as their Preference, while 34% and 21% have Economy and Need, consequently as their reason. More than half of users in Cluster 3 opts for MRT as their Preference. Another 18% uses MRT for its economic reason, and the remainder uses MRT as a necessity. Respondents in Clusters 4 and 5 have a similar answering pattern. Most of them chose MRT because of their preferences followed by Economy, and a tiny percentage of respondents use MRT because they need to.

Lastly, the ranking for the frequency of using MRT is similar in most clusters. A larger proportion of respondents in each cluster only uses the MRT Once in a while. During weekdays, where people usually commute to work, 13% of Cluster 2, 14% of Cluster 3 and 4 and, 5% of Cluster 5 respondents use the MRT. As for the weekends, only a small percentage of respondents use the MRT, with one (1) person in Cluster 1, Cluster 3, and Cluster 4. The results show that respondents hardly use the MRT as their first option to commute (Table 3).

Table 3 Travel pattern of respondents

Question	Responses	Below 20 (C1)		21–35 (C2)		36–50 (C3)		51–60 (C4)		Above 60 (C5)	
		N	%	N	%	N	%	N	%	N	%
Trip purpose	– To or from work	0	0	19	36	8	36	2	4	4	18
	– Shopping/Eating out	0	0	22	42	5	23	12	32	5	23
	– Recreation/Social/Place of worship	1	50	6	11	7	32	12	32	7	32
	– School	1	50	4	8	0	0	0	0	0	0
	– Medical/Dental	0	0	0	0	0	0	0	0	0	0
	– Other	0	0	2	4	2	9	12	32	6	27
Reason of using MRT	– Need	0	0	11	21	3	14	3	8	1	5
	– Economy	1	50	18	34	4	18	11	29	4	18
	– Preference	1	50	24	45	15	68	24	63	17	77
Travel frequency	– Once in a while	1	50	27	51	15	68	28	74	12	54
	– Occasionally	0	0	17	32	2	9	5	13	5	23
	– Weekends	0	0	1	2	2	9	2	5	2	9
	– Weekdays	0	0	7	13	3	14	2	5	3	14
	– Every day	1	50	1	2	0	0	1	3	0	0

* N = Frequency, % = Percentage

4.3 Importance-Performance Analysis (IPA) of Each Cluster

The IPA results are tabulated and graphed for each cluster, which is separated by age range. Cluster 2 was used as an example because this age group has the most significant number of respondents.

4.3.1 21–35 (Cluster 2)

Individuals in this category has the greatest number of respondents. Table 4 indicates the average performance and importance as well as the gap between those two factors for each dimension, while Fig. 2 shows the IPA results for Cluster 2. For this set of respondents, empathy and safety and security dimension is located in quadrant A "Keep up the good work", meaning that they are very satisfied with the performance as weighing to its importance. Meanwhile, there are two dimension that lie in quadrant C "Low priority", namely assurance and responsiveness. This demonstrates that most customers are unhappy with these characteristics but do not consider they are significant. These qualities also need to be improved, but only after the qualities in quadrant B have been improved (concentrate here), where in this case the dimension accessibility is highlighted.

Table 4 Mean performance and mean importance of Cluster 2

Attributes	Mean performance	Mean importance	Gap
Reliability	4.12	4.37	−0.25
Assurance	3.78	4.31	−0.53
Tangibles	4.03	4.36	−0.34
Empathy	4.22	4.41	−0.18
Responsiveness	3.66	4.25	−0.59
Safety & Security	4.08	4.48	−0.41
Accessibility	3.97	4.46	−0.49
Total	3.98	4.38	–

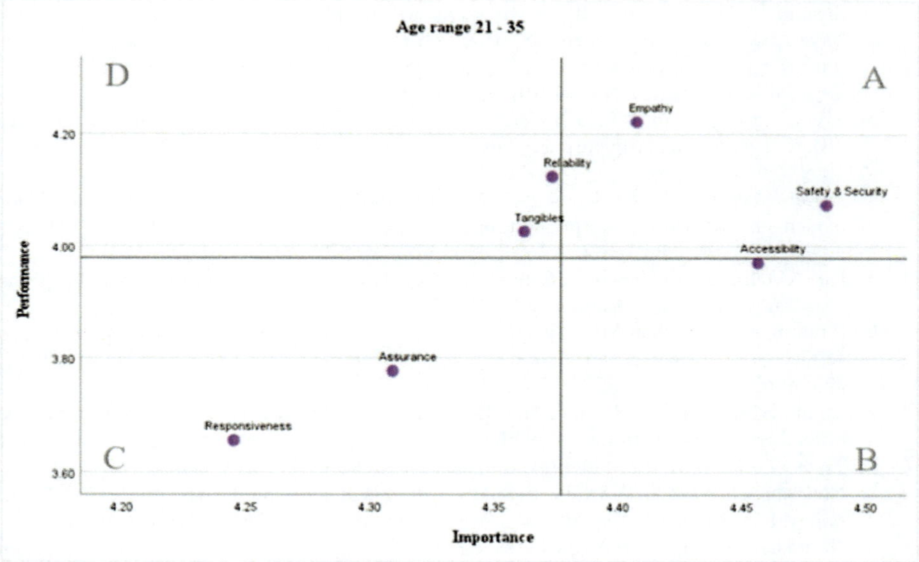

Fig. 2 IPA results for cluster 2

5 Conclusion

The objective was to acquire information on respondents' satisfaction in MRT services by segmenting them according to their age groups. It is revealed that the dimension that affects the perception of customers varied significantly. This objective contradicts the study made by C.Morton [21]. The results are shown in this study that the young (35 and under) and elderly generation (above 60) perceived the tangibility dimension has the most negligible impact on attracting customers. Nevertheless, the mid-range age group (36–60) viewed the tangibility dimension as vital and good performance. Despite the variation, some dimensions have a similar perspective in all ages. For example, most age groups are satisfied with the factors in the safety and security dimension. Therefore, it is suggested that stakeholders and authorities maintain or expand the budget allocated in the safety and security dimension.

Acknowledgements The first author would like to thank UNITEN BOLD 2021 Grant J510050002/2021167 for making it possible to publish this article.

References

1. Ibrahim ANH et al (2021, January) Gender and age do matter: exploring the effect of passengers' gender and age on the perception of light rail transit service quality in Kuala Lumpur, Malaysia. Sustain 13(2):1–18. https://doi.org/10.3390/su13020990
2. Yaya LHP, Fortià MF, Canals CS, Marimon F (2015) Service quality assessment of public transport and the implication role of demographic characteristics. Public Transp 7(3):409–428. https://doi.org/10.1007/s12469-014-0099-7
3. Sze-siong C, Aksan A (2018) Users' satisfaction with mass rapid transit and the effect of different genders and trip purposes. Int J Acad Res Bus Soc Sci 8(15):131–145. https://doi.org/10.6007/IJARBSS/v8-i15/5097
4. Zefreh MM, Hussain B, Sipos T (2020, July) In-depth analysis and model development of passenger satisfaction with public transportation. KSCE J Civ Eng 2020 2410 24(10):3064–3073. Doi: https://doi.org/10.1007/S12205-020-1871-7
5. Das AM, Ladin MA, Ismail A, Rahmat ROK (2013) Consumers satisfaction of public transport monorail user in Kuala Lumpur. J Eng Sci Technol 8(3):272–283
6. Ibrahim ANH, Borhan MN, Yusoff NIM, Ismail A (2020, May) Rail-based public transport service quality and user satisfaction – a literature review. Promet-Traffic & Transp 32(3):423–435. https://doi.org/10.7307/PTT.V32I3.3270
7. Irtema HIM, Ismail A, Borhan MN, Das AM, Alshetwi ABZ (2018, December) Case study of the behavioural intentions of public transportation passengers in Kuala Lumpur. Case Stud Transp Policy 6(4):462–474. https://doi.org/10.1016/J.CSTP.2018.05.007
8. Norhisham S et al (2020) Evaluating passenger load factor of public bus services in West Klang Valley. Lect Notes Civ Eng 59:95–102. https://doi.org/10.1007/978-981-15-1193-6_11
9. Norhisham S et al (2019) Assessing stakeholder's view forquality bus servicesin Kuala Lumpur. Int J Adv Sci Technol 28(10):359–369
10. More than 100 delays on Kajang MRT Line | The Star. https://www.thestar.com.my/news/nation/2020/07/15/more-than-100-delays-on-kajang-mrt-line. Accessed: October 07, 2020.
11. Six MRT stations hotspots for crime, Prasarana taking steps to address problem | The Star. https://www.thestar.com.my/news/nation/2019/02/20/six-mrt-stations-hotspots-for-crime-prasarana-taking-steps-to-address-problem/. Accessed: October 07, 2020.
12. Thanaraju P, Khan PAM, Sivanathan S, Juhari NH (2019) Passengers' satisfaction towards railway facilities (RAILQUAL in the central region). Int J Recent Technol Eng 8(2):561–571 (Special Issue)
13. Esmailpour J, Aghabayk K, Abrari Vajari M, De Gruyter C Importance – Performance Analysis (IPA) of bus service attributes: a case study in a developing country. Transp Res Part A Policy Pract 142(November):129–150. Doi: https://doi.org/10.1016/j.tra.2020.10.020
14. Coppola P, Silvestri F (2020) Assessing travelers' safety and security perception in railway stations. Case Stud Transp Policy 8(4):1127–1136. https://doi.org/10.1016/j.cstp.2020.05.006
15. Mat A et al (2018) The influence of public commuters satisfaction towards Mass Rapid Transportation (MRT) Services among Kuala Lumpur riders: a PLS-SEM approach. (IJEKM) Int J Educ Knowl Manag Fac Bus. Manag 1(1):1–11. Accessed: June 30, 2021. [Online]. https://rpajournals.com/ijekmJournalHomepage, https://rpajournals.com/ijekm

16. Lai IKW, Hitchcock M (2015, June) Importance-performance analysis in tourism: a framework for researchers. Tour Manag 48:242–267. https://doi.org/10.1016/j.tourman.2014.11.008
17. Revindran M, Ragen PNK, Mahmud B (2020, April) A study on logistics service quality in e-retailing amongst online shoppers in Kuala Lumpur. IOP Conf Ser: Mater Sci Eng 780(6). Doi: https://doi.org/10.1088/1757-899X/780/6/062016
18. Norhisham S et al (2019) Statistical overview on quality bus services in Klang Valley. Int J Adv Sci Technol 28(10):370–380
19. Lwesya F, Jaffu R (2017) Customer service quality management in public transport: the case of rail transport in Tanzania. Int Rev 3–4:102–117. https://doi.org/10.5937/intrev1704102l
20. Ibrahim, Borhan, Zakaria, and Zainal (2019) Effectiveness of commuter rail service toward passenger's satisfaction: a case study from Kuala Lumpur, Malaysia. Int J Eng Technol 8(1.2):50–55. Doi: https://doi.org/10.14419/ijet.v8i1.2.24871
21. Morton C, Caulfield B, Anable J (2016) Customer perceptions of quality of service in public transport: evidence for bus transit in Scotland. Case Stud Transp Policy 4(3):199–207. https://doi.org/10.1016/j.cstp.2016.03.002

The Role of Geotechnical Engineers in Sustainable Construction Processes: A Regard to Soil-Structure Interaction Problems

Andrea Galli

Abstract Sustainability-related issues became more and more important in these last decades owing to the increasing awareness in modern societies of the limitedness of the resources at the global planetary scale. Environmental resources, in particular, play a key role in this perspective since (i) they are hardly recoverable and (ii) their uncontrolled consumption may trigger complex chains of events, with potentially catastrophic consequences. In technological advanced societies like ours, the Engineering contribution is essential to correctly deal with these problems, and engineers (both scientists and practitioners) cannot disregard these topics. With reference to Built Environment, the contribution of Geotechnical Engineers can in particular give relevant positive benefits, since Geotechnical works very often involve large (or very large) soil volumes, and are in general conceived as long-lasting (if not permanent) interventions. In the paper, with specific reference to soil-structure interaction problems, some general considerations are presented, by putting in evidence the contribution that Geotechnical Engineers can give to the design process, since its early stages. Several examples of recent case studies will be reviewed, with reference to the protection of historic and cultural heritage, to slope stabilizing interventions, to the protection of strategic infrastructures like buried pipelines, in particular by employing innovative upscaling approaches based on the "macroelement" concept.

Keywords Sustainable constructions · Geotechnical works · Macroelement model · Soil-structure interaction · Monitoring data

1 Introduction

Sustainability of anthropic interventions (or, more in general, of human impacts on environment and society) is a very complex topic, involving not only technical issues, but also ethical and social themes. In the last forty years, an increasing interest on sustainability has been observed, especially towards environmental-related issues.

A. Galli (✉)
Department of Civil and Environmental Engineering, Politecnico di Milano, Milan, Italy
e-mail: andrea.galli@polimi.it

E. M. Nia et al. (eds.), *Sustainable Development Approaches*, Lecture Notes in Civil Engineering 243, https://doi.org/10.1007/978-3-030-99979-7_7

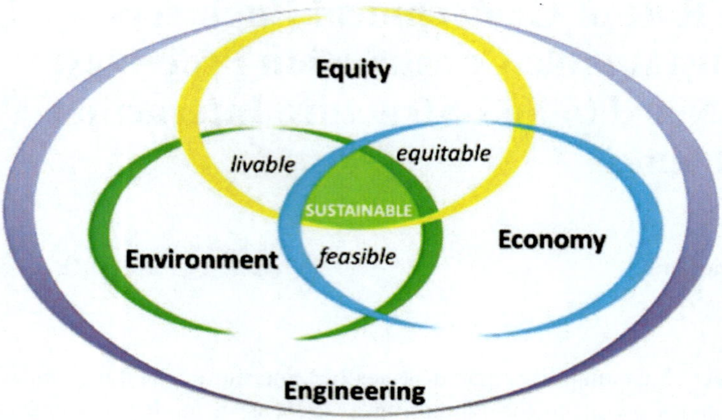

Fig. 1 The "four Es" of sustainability in engineering projects

This is due to the increased awareness of modern societies of the limitedness of the available resources (they are in general hardly recoverable) and to the fact that even small modifications in some fundamental parameters (e.g. average Earth temperature, due to climate change) can trigger unexpected—and sometimes catastrophic—chains of events. Early definitions describe a sustainable system as a system "able to satisfy its needs without diminishing the chance of future generations" [1], and a sustainable development as the one that "meets the needs of the present without compromising the ability of future generations to meet their own needs" [2]. Both of these definitions implicitly introduce some fundamental concepts such as social and intergenerational equity over time.

Sustainability, however, is not only a matter of correctly balancing over time the resources (i.e. the capacities of a system) and the demands (i.e. the loads acting on it), but it is rather related to finding the right compromise among the so-called "four Es" of sustainability, i.e. Equity, Environment, Economy and, since we live in technologically advanced society, Engineering ([3]; Fig. 1). Sustainability, moreover, often implies also resilience issues, i.e. the capability of a system to recover its functionality (fully or partially) after a disruptive event. Sustainability necessarily deals then with multidimensional aspects sometimes in conflict with each other, in a context of scarcity of information, and in a long-lasting perspective. From a theoretical point of view, the search for a sustainable solution in design problems is often a "wicked problem", i.e. "a class of social system problems which are ill formulated where the information is confusing, where there are many clients and decision makers with conflicting values, and where the ramifications in the whole system are thoroughly confusing" [4]. As a consequence, sustainable design solutions can hardly be objectively found only on the basis of quantitative metrics (e.g. based on environmental data, like global warming potential or carbon footprint; see e.g. [5, 6]) or qualitative indicator systems (as GeoSPeAR, developed by [7, 8]). They rather need to be based on rational judgments, in a holistic approach capable of balancing the four Es and of sharing the information among all the stakeholders. Detailed investigations of such aspects are evidently out of the purposes of the present paper; interested readers can however find a fundamental review in [3].

Without any claim of completeness, a possible formal framework helping in highlighting the main aspects involved in the design of sustainable interventions can be derived as follows. Let assume that the system under consideration is at a state F, depending, through a multidimensional function f, on a vector α of "natural actions" (e.g., seasonal water table oscillations, or earthquake activity), on a vector β of "anthropic actions" (e.g., the realization of a new infrastructure) and on a set κ of other variables (in general depending on the scale of the analysis, e.g. the single building, the infrastructure, the considered municipality or region, the whole planet, ...). Both α and β may vary with time t, so that:

$$F = f(\alpha(t), \beta(t), \kappa) \tag{1}$$

Sustainability requirements implies that this global state F of the system does not change with time, or at least that its rate of variation does not exceed a (small) quantity ε, representing a sort of "allowable consumption rate"

$$0 \leq \frac{dF}{dt} = \frac{\partial f}{\partial \alpha} \cdot \frac{d\alpha}{dt} + \frac{\partial f}{\partial \beta} \cdot \frac{d\beta}{dt} \leq \varepsilon \tag{2}$$

If for example function f is assumed to be represented by the entropy of the system, then Eq. (2) reflects the Second Law of Thermodynamics. Equation (2) allows to highlight the following four quantities:

- the term $\frac{\partial f}{\partial \alpha}$ represents the effects of natural actions on the system, and it then falls within the usual research domains of Natural Sciences, Environmental Engineering and/or Social Sciences;
- the rate $\frac{d\alpha}{dt}$ measures the intensity of natural actions, which is typically the object of field monitoring activities;
- the term $\frac{\partial f}{\partial \beta}$ expresses the effects of anthropic actions on the system, and it is then usually related to industrial or construction processes, to technological and scientific skills, or, more in general to social and economic processes.
- $\frac{d\beta}{dt}$, expressing the anthropic actions (i.e. the design choices), resulting from the adopted governing policies or decision-making processes.

By solving the inequalities expressed in Eq. (2), two conceptual limitations can be explicitly found for the anthropic actions $d\beta/dt$, helping in defining a sort of "fan" of sustainable actions

$$-\left(\frac{\partial f}{\partial \beta}\right)^{-1} \cdot \frac{\partial f}{\partial \alpha} \cdot \frac{d\alpha}{dt} \leq \frac{d\beta}{dt} \leq \left(\frac{\partial f}{\partial \beta}\right)^{-1} \left(\varepsilon - \frac{\partial f}{\partial \alpha} \cdot \frac{d\alpha}{dt}\right) \tag{3}$$

The graphical interpretation of Eq. (3) is sketched in Fig. 2a, where four different actions (labelled a, b, c, and d, respectively) are shown, starting from the present state β_0 (i.e. the present asset of constructions in the built environment, for example)

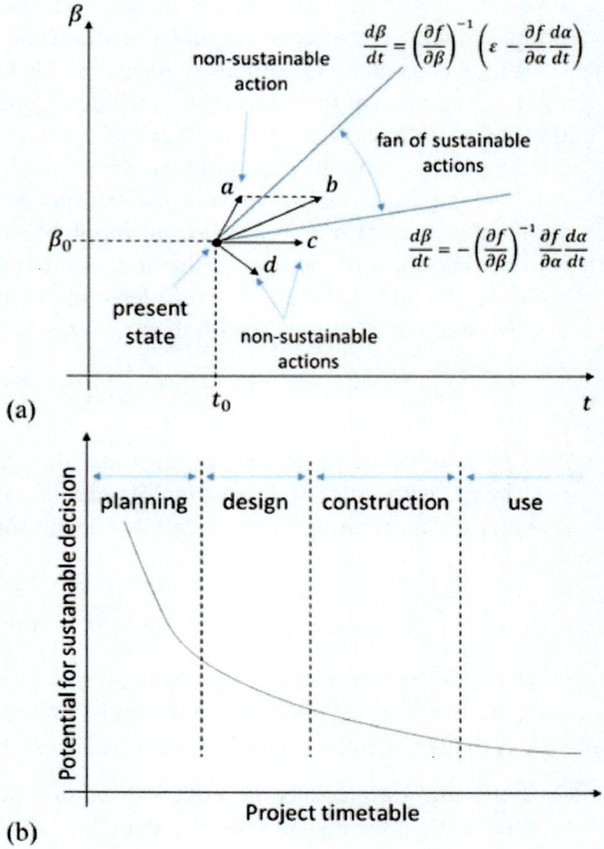

Fig. 2 **a** conceptual definition of sustainable actions; **b** typical steps in geotechnical projects (modified from [3])

at the current time t_0. Three of them (a, b, and d) fall outside the fan of directions defined in Eq. (3) and represent non-sustainable actions, whilst only action b, in this example, respects the sustainability criteria. Such conceptual visualization helps also in clarifying the importance of the rapidity of implementation of each action, since in some cases the same intervention may fall inside or outside of the sustainable fan (like actions b and a, respectively). It also shows that apparently acceptable actions (e.g. actions c, corresponding with the "doing-nothing" option, having $d\beta/dt = 0$, or d, exemplifying a decommissioning case) may in principle represent non-sustainable options for the case study under consideration.

For the sake of completeness, it should finally be observed that even in the ideal case of "perfectly" sustainable actions (i.e. zero allowed consumption rate, with $\varepsilon = 0$), for which the fan of sustainability degenerates to one single path, multiple solutions can be found, since matrix $\partial f/\partial\beta$ may in general contain alternative and competing technologies. This analysis, although simplified, clearly highlights the unavoidable need of multidisciplinary approaches (owing to the combination of the quantities $\partial f/\partial \alpha$ and $\partial f/\partial \beta$, representing the interaction among different scientific domains) and the essential role of monitoring systems ($d\alpha/dt$) in the correct statement of a sustainability problem.

Built Environment and Civil Engineering are among the most important fields of application of sustainability-related criteria, and Geotechnical Engineering is particularly involved in both them. Geotechnical interventions, in fact, usually involve large (or even very large) volume of soils, like for example the case of infrastructural systems or slope stabilizing interventions, and are in general conceived as long-lasting (or even permanent) works. [9, 10] identified among others a wide list of possible topics for sustainable geotechnical engineering, like e.g. the use of environment-friendly materials; the use of energy-efficient geotechnical techniques for site investigation; the reuse of foundations and other geotechnical structures; the use of underground space; the development of geotechnical techniques for pollution control and redevelopment of brownfields; the mitigation of geohazards (like e.g. landslides). In the aforementioned work [3], three recent trends are nowadays recognized as possible ways to incorporate sustainability issues in Geotechnical projects [11]:

(i) geostructures should be designed for performance rather than for ease of construction;
(ii) site-specific requirements need to be considered in the design process;
(iii) designers have to consider soil–structure interaction problems rather than to separately analyse structural and foundation parts.

Of course, the earlier the geotechnical issues are considered in the design process, the more effective the final design solution will be in term of achieving sustainability requirements (Fig. 2b). For example, with reference to the design of a stabilizing intervention for an unstable slope, incorporating geotechnical issues during the initial planning phase would allow to optimize the position as well as the choice of the intervention typology (e.g. to choose among retaining wall, deep soil nailing, stabilizing piles, basal embankment, …). During the subsequent design phase, less additional choices will only be available for the designer, as for example the type of construction materials or some minor dimensioning details. During the construction phase, only limited technological aspects can generally still be modified and finally, after the construction phase, no remarkable benefits can in general be achieved during the standard use of the structure along its whole lifespan.

In the present paper, some cases of soil-structure interaction problems will be considered, in particular by adopting a generalized upscaling approach based on the Macroelement concept. If diffusely adopted in the design practice, at Author's feeling this could potentially represent a valid and economically convenient example of Scientific Transfer in Geotechnical Engineering. Recent cases studies regarding protection of historic and cultural heritage, slope stabilizing structures, protection of strategic infrastructures will be discussed, with the aim of highlighting the importance of a "geotechnically-sound" approach for matching the aforementioned fan of sustainable actions.

2 The Macroelement Concept

The design of Geotechnical interventions are very often characterized by a high complexity degree (due to the non-linear and irreversible mechanical behaviour of the soil, to thermo/hydro/chemo coupling effects, to the complex geometry of real boundary value problems, ...), in a context of scarcity of information (especially with respect to soil mechanical characterization, and/or to the spatial distribution and time variability of the applied environmental loads). Accurate modelling of such problems would in general require the adoption of the most suitable constitutive rules at the local scale of the Representative Element Volume of soil (REV), and its integration over the entire domain, once equilibrium and compatibility equations are imposed (an interesting work on the importance of the adopted constitutive rule for soil with respect to sustainability issues in Geotechnical problems was recently published by [12]). This process is in general done by means of numerical codes (e.g. based on Finite Element or Finite Difference approaches) discretizing the entire soil domain in a number of elements of finite size and, in case of time-evolving problems, by step-wise incrementally solving the governing equations. In this perspective, Soil-Structure Interaction problems (SSI) are recognized as inherently multiscale problems, since they need to deal with the mechanical behaviour of the system at the local scale of the REV, but with the aim of reproducing the behaviour at the macroscale of the entire structure (e.g. the settlement of a retaining wall, or of a direct foundation), or even to include it into a slope stability problem at the megascale of the entire slope. Even if technically feasible, such multiscale complexity can hardly be tackled by standard commercial numerical codes usually available for design applications. Advanced theoretical skills are required (in particular with respect to the choice and calibration of the adopted constitutive parameters) together with large computational resources, so that such numerical methods cannot yet be considered as standard and user-friendly tools, in particular when preliminary design phases are considered, where a large amount of parametrical analyses are in general required in order to identify a preliminary set of possible alternative design solutions. This is particularly detrimental with respect to sustainability criteria, since the initial planning phase is often highly effective in increasing the overall sustainability of the project (see Fig. 2b).

In this perspective, Macroelement approaches may represent a possible alternative modelling option, since they allow to describe SSI in terms of the generalized stress and strain variables, representing the resulting load components (vertical, V, and horizontal, H) or moments (M) acting on the system and of the corresponding pattern of structural displacements (vertical, v, and horizontal, u) or rotations (θ). Macroelement approaches allow then to upscale the problem from the local scale of each REV to the macro scale of the structure by introducing an ad hoc "generalized" incremental constitutive relationships, between the abovementioned generalized stress and strain variables. The structure and the soil interacting with it are then considered as a unique "macro"element, whose behaviour is fully described by means of only few degrees of freedom (Fig. 3a).

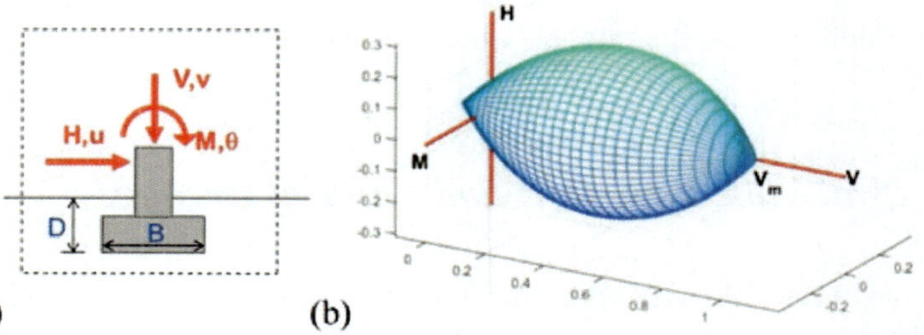

Fig. 3 **a** definition of static (V, H, M) and kinematic (v, u, θ) quantities for a macroelement for a shallow strip foundation, with a width B and at a depth D below the ground surface; **b** example of the failure condition for a macroelement in a generalized stress space

An original version of a macroelement model for small scale shallow strip foundations on sand was proposed by [13, 14], by assuming a rigid-plastic relationship between the vectors of load increments $\dot{\boldsymbol{Q}} = \left[\dot{V}\ \dot{H}\ \dot{M} \right]^T$ (i.e. the generalized stress), and displacement increments $\dot{\boldsymbol{q}} = \left[\dot{v}\ \dot{u}\ \dot{\theta} \right]^T$ (i.e. the generalized strain). A coupled failure condition was assumed in the (V, H, M) loading space (Fig. 3b), together with a yielding surface obeying an isotropic strain hardening law. By adopting standard plasticity approaches, plastic (i.e. permanent) foundation settlements can rapidly be computed under combined non-monotonic loading paths, once a plastic potential is introduced and the consistency condition is imposed to the yielding surface. Since this pioneering version of the macroelement, several evolutions have been proposed in the last thirty years, in particular regarding the shape of the yielding surface [15, 16], or by generalizing to 3D geometrical conditions for isolated foundations [17–24]. More recently, advanced constitutive hypotheses have been included into the macroelement formulations in order to reproduce cyclic behavior [25, 26], rock boulder impacts on granular layers [27], or to study dynamic and seismic problems [28–31]. Some hypoplastic versions of the macroelement have also been proposed in this last decade (see e.g. [32]), by introducing the concept of "internal displacement", mimicking the "intergranular strain" originally defined by [33] at the REV scale. For interested readers, a recent critical review on the application of macroelement concepts in Geotechnical engineering is presented in [34].

3 Case Studies

In this section, some examples of application of the "macroelement" concept to real case studies will be briefly reviewed. In particular, a case study regarding the protection of historic and cultural heritage in northern Italy, a case study on landslide risk mitigation for a railway transportation system, and some considerations about the protection of strategic infrastructures like oil and gas pipelines will be discussed, thus highlighting the theoretical concepts concisely presented in the Introduction.

3.1 Protection of Historic and Cultural Heritage

The town of Lecco is located in Northern Italy and is built on the Adda river (Fig. 4a), with three bridges connecting the two sides of the river. The most ancient one is the Azzone Visconti bridge, built in 1336–1338 and constituted by 11 masonry arches, with a total length of 132.6 m, an average height of about 10 m over the river, and with the largest arch measuring 14.6 m (Fig. 4b). It is the most ancient masonry structure in the town, and one of the most important historical bridges in the region (some recent studies suggest that this bridge is a part of the background of the famous Mona Lisa masterpiece of Leonardo da Vinci; see [35]). In 2013–2015 the bridge was subject to a campaign of experimental investigations, with both on site and laboratory tests, with the aim of evaluating its ultimate bearing capacity with respect to the increased traffic loads, and to perform site-specific seismic analyses (detailed descriptions in [36–39]).

The foundation of the masonry piers of the bridge were originally constituted by wooden piles, driven into the riverbed. In 1949–1950 some important strengthening works were done, consisting in encapsulating the masonry piers with a steel sheetpile forming a closed ring with a depth of about 5 m below the river bed (Fig. 5a), finally bounded with a R/C continuous ring (Fig. 5b). These interventions were aimed at confining the soil immediately below the pier in a pseudo-oedometric condition, thus increasing the depth of activation of possible failure mechanisms in the soil at the base of the pier (i.e. its burying depth) and increasing the ultimate bearing capacity of the foundation system.

For the purposes of the present paper, the seismic analyses discussed in [38], obtained by means numerical 3D dynamic FEM analyses, will only be considered. The results have been interpreted in the light of the interaction domain of a macroelement, thus allowing a rapid evaluation of the effects of the strengthening works on the overall behaviour of the foundations. Seven site-specific synthetic accelerograms have been considered in the numerical analyses (for the sake of completeness, Fig. 6a shows one of the selected accelerograms) and the results are summarized in Fig. 6b and c in terms of the loading paths computed for piers 1, 5, 6, and 10, in the $V - H$ plane, normalized with respect to the vertical bearing capacity V_m of each foundation, respectively. In the two plots, the trace of the normalized failure locus is also reported.

The results show how the strengthening interventions have markedly increased the average values of the static safety factors of the piers with respect to the vertical load only. For piers 5 and 6, in particular, the values of the ratios V/V_m are reduced, owing to the increase in V_m, from 0.45 to 0.1, implying an increase in the safety factor from 2.2 to about 10. Nevertheless, the safety with respect to possible failure due to horizontal load has not been significantly modified, since in both cases the

Fig. 4 **a** position of the town of Lecco in Italy and panoramic view of the lake of Lecco and of the location of the Azzone Visconti bridge; **b** view of the Azzone Visconti bridge in 2014, at the beginning of the investigation campaign

loading paths show approximately the same maximum obliquity (i.e.the maximum ratio H/V) and the same distance from the corresponding failure value H_m of the horizontal loading component. At the time of the intervention, of course, the concept of the macrolement was not yet available for the engineers, and seismic analyses were not so diffuse (it must also be noted that the town of Lecco is in a low seismicity zone). Nevertheless, nowadays, the dimensioning of this strengthening intervention would not be considered fully effective, since it improves the safety of the structure

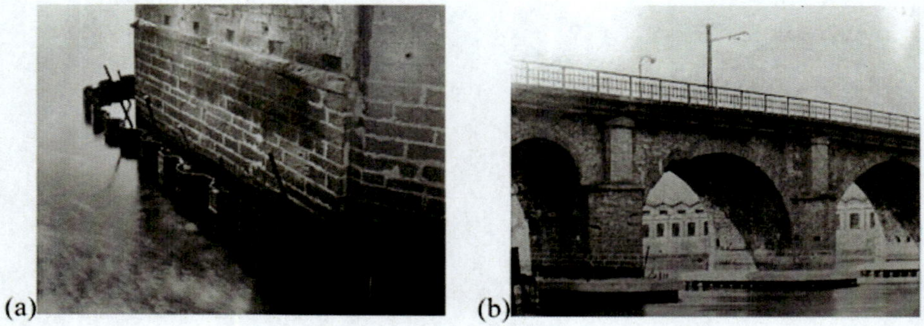

Fig. 5 Strengthening work executed in 1949–1950; **a** installation of the steel sheetpile encapsulating the masonry pier; **b** positioning of final cast in place the R/C ring. Both images are courtesy of Consorzio dell'Adda

with reference only to a limited class of failure mechanisms (i.e. the vertical bearing capacity), disregarding those related to horizontal loading components. These latter are however responsible for possible sliding and rocking of the foundation, with important torsional loads transferred to the bridge deck.

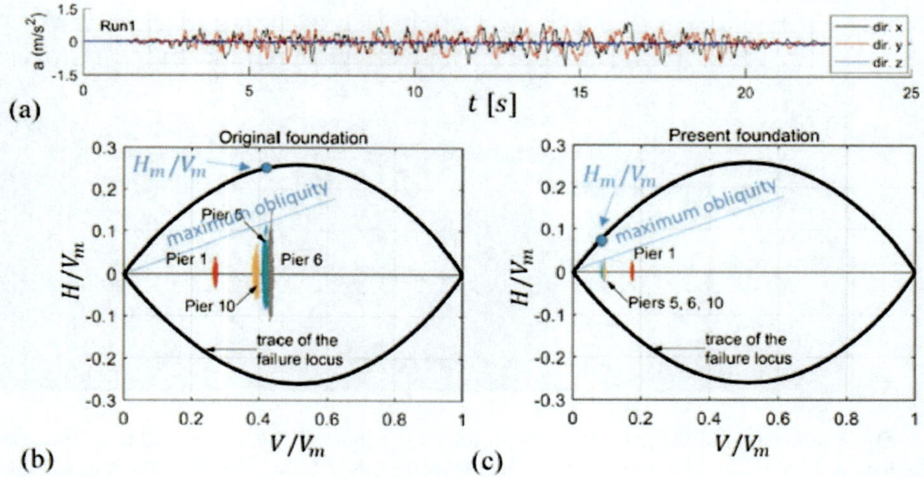

Fig. 6 **a** example of the syntethic accelerogram used in the numerical analyses; seismic loading paths for the vertical and horizontal components for piers 1, 5, 6 and 10 in the normalized loading space for **b** the original and **c** the present foundation (modified from [38])

3.2 Landslide Risk Mitigation

If in the previous example the focus was on the SSI problems (and then, with reference to the definitions listed in the Introduction, on the term $\partial f/\partial \beta$), hereafter the focus will on the contrary be on the monitoring issues (i.e. the term $d\alpha/dt$). The case study refers to slope stabilizing piles, for which the SSI problem is simplified by introducing ad hoc non-linear springs to reproduce the local interaction between the pile and the soil (Fig. 7a; [40, 41]). By imposing, along the depth z below the ground level, a profile $U(z)$ to soil displacement over the thickness H of the unstable soil layer, and by stepwise integrating the non-linear mechanical response of the springs, it is possible to derive a "characteristic curve" representing the evolution of the stabilizing action A provided by the pile as a function of the observed soil displacement pattern. When adopting the so called "hybrid methods" ([42]), this function can be used in the standard stability analyses in order to derive the usual value of the factor of safety of the slope as a function of the soil deformation level. When adopting full displacement-based methods ([41]), this characteristic curve can again be introduced into the motion equation of the unstable soil mass, but with the aim to directly compute the expected soil displacement, and its time evolution. This is particularly important for quantitatively assessing the efficiency of the intervention in terms of (i) reduction of the landslide displacement rate, (ii) amplitude U_d of the final expected soil displacement and (iii) delay t_d, after the installation time t_0, needed to reach the final stable condition (Fig. 7b).

This approach has been employed to the preliminary design of anchored stabilizing piles for a railway line, built on a steep slope (about 40° to 42°) with a total height of 15 m (Fig. 8a). The slope below the railway ballast is composed of alternate layers of fine cohesive materials and highly fractured weakly cemented rocks, and slow downward settlements have been measured in the slope by means of a long term inclinometer monitoring system (interested readers can find further details in [43]). The readings (Fig. 8b) showed that slow soil movements, distributed along

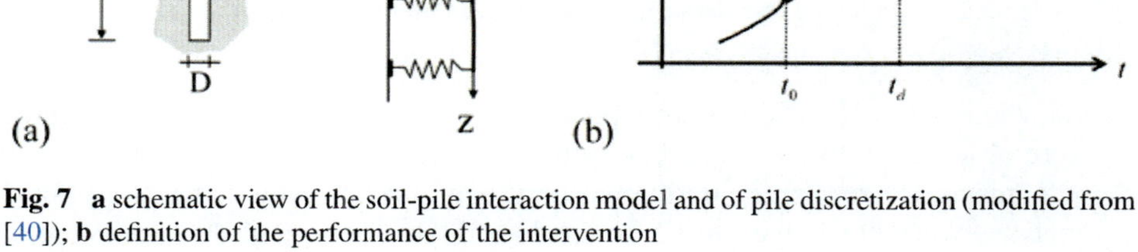

Fig. 7 **a** schematic view of the soil-pile interaction model and of pile discretization (modified from [40]); **b** definition of the performance of the intervention

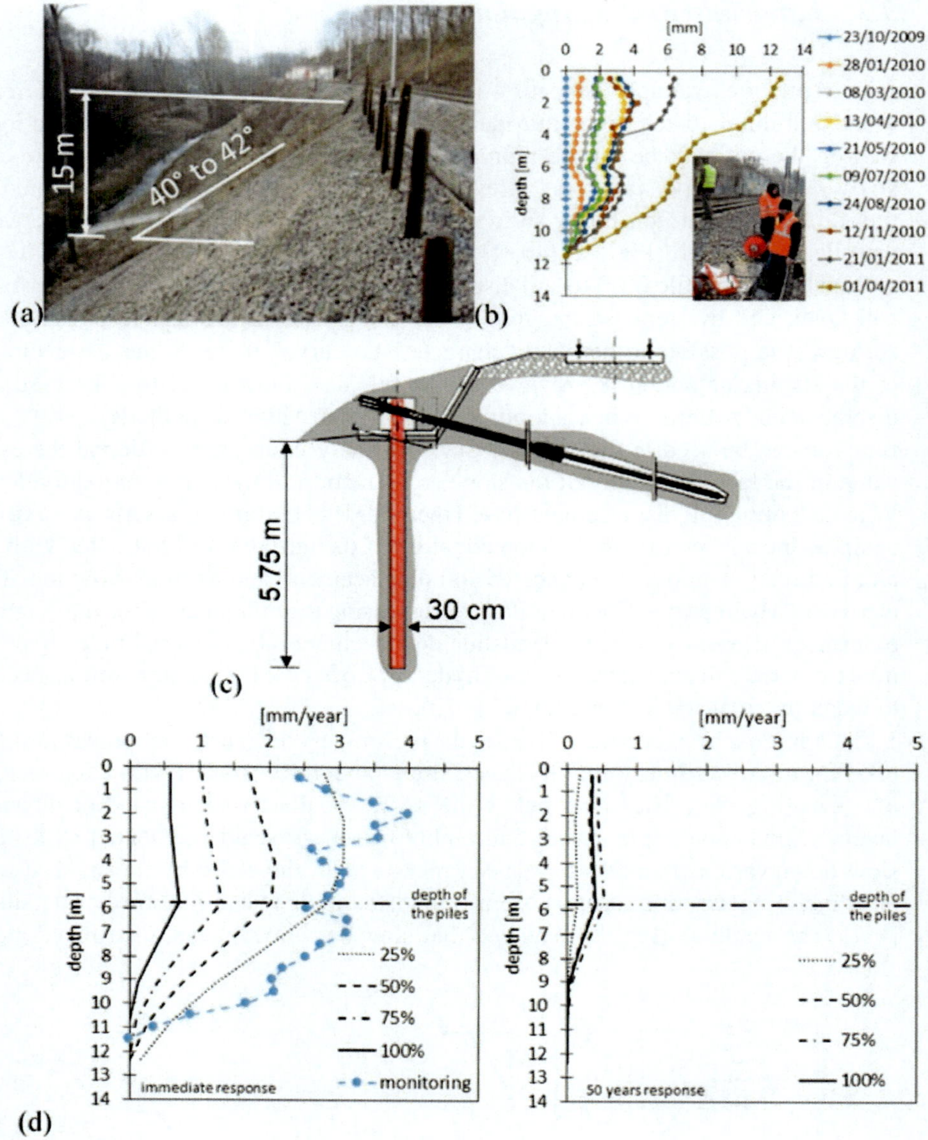

Fig. 8 **a** view of the embankment and of the slope; **b** inclinometer installation and partial report of the monitored data; **c** sketch of the anchored micropile; **d** expected profiles of soil displacement rate immediately after the installation and after 50 years (modified from [43])

the whole depth of the slope, were presents. No active and localized failure surfaces were observed in to the slope, but, rather, a diffused soil displacement pattern was recognized. Consequently, traditional limit equilibrium methods for slope stability analyses were not judged to be applicable for this case, and the aforementioned displacement-based methods were implemented.

The proposed structural intervention consisted in the realization of micropiles, 5.75 m long (i.e. less than a half of the slope height) and with a diameter of 30 cm (Fig. 8c), installed with a spacing-to-diameter ratio equal to 2, and anchored with partially prestressed ground anchors below the railway. The displacement-based approach allowed to parametrically design the structure, particularly focusing on the optimization of structural dimensions and on the desired performance of the intervention. The project requirements were in fact to reduce the slope displacement rate below an allowable threshold value of 0.5 mm/year, without inducing significant heave of the railway when imposing the initial prestressing action to the ground anchors. Several scenarios were then numerically taken into consideration, by imposing 25, 50, 75 or 100% of anchors ultimate tensile capacity, and the results were expressed in terms of the expected profile of soil displacement rates immediately after the installation and at the end of the lifespan (50 years) of the interventions (Fig. 8d). By balancing the requirements, the best installation option was to impose a prestressing action of about 75% of the anchor capacity, in order to immediately reduce as much as possible the soil displacement rate (to about 1 mm/year, when the monitoring system showed that current values were 3 to 4 mm/year), without inducing excessive uphill displacement to the slope. In long term condition however, all the investigated scenarios give a residual soil displacement rate below the required threshold of 0.5 mm/year.

3.3 Protection of Strategic Infrastructures: Buried Pipelines

A third case study is finally reported hereafter, regarding again a particular SSI problem, and involving the protection of buried pipeline against slow soil movements, as those characterizing creeping landslides. In this case the displacement-based approach is not employed to efficiently compute the global stabilizing action that a structure can give to an unstable slope (pipelines cannot be considered as stabilizing structures), but to accurately evaluate the additional stresses that a creeping landslide induces in buried pipelines.

The idea is again to model the SSI by introducing an array of coupled 3D macroelement, reproducing the local interaction between the soil and the pipeline, and modeling the landslide by imposing a known soil displacement pattern $\boldsymbol{U}$ at the base of the macroelements (Fig. 9). [44, 45] presented a simplified piece-wise linear technique for the numerical integration of the macroelement law, here specifically conceived for buried pipelines [46], and thus allowing to run accurate displacement based analyses in few tens of seconds (on a standard laptop).

Figure 10 shows an example of application of such technique to a real case study of a steel pipeline (about 1600 m long; diameter of 75 cm), buried at a depth of about 2 m along the side of a hill. The soil is characterized as a fine poorly cohesive material, with unit weight of 19 kN/m^3, friction angle of 20° and cohesion of 10 kPa, equivalent Young modulus of 8 MPa and Poisson coefficient of 0.3. The active landslide area is diffused along the pipeline (Fig. 10a), with a dominant axial

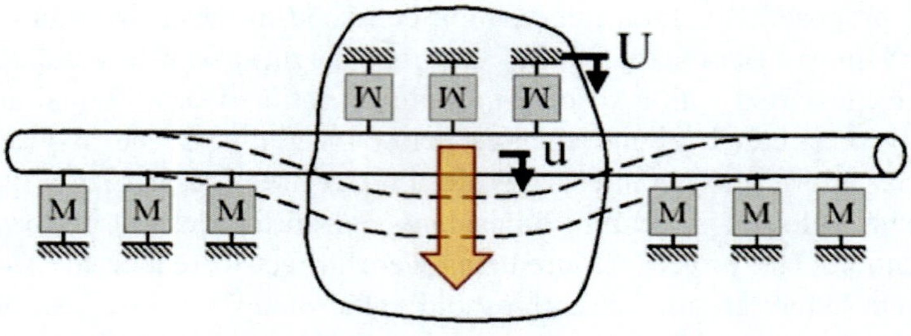

Fig. 9 Schematic view of a buried pipeline, interacting with soil by means of a series of 3D macroelement and subject to a soil displacement pattern ***U***

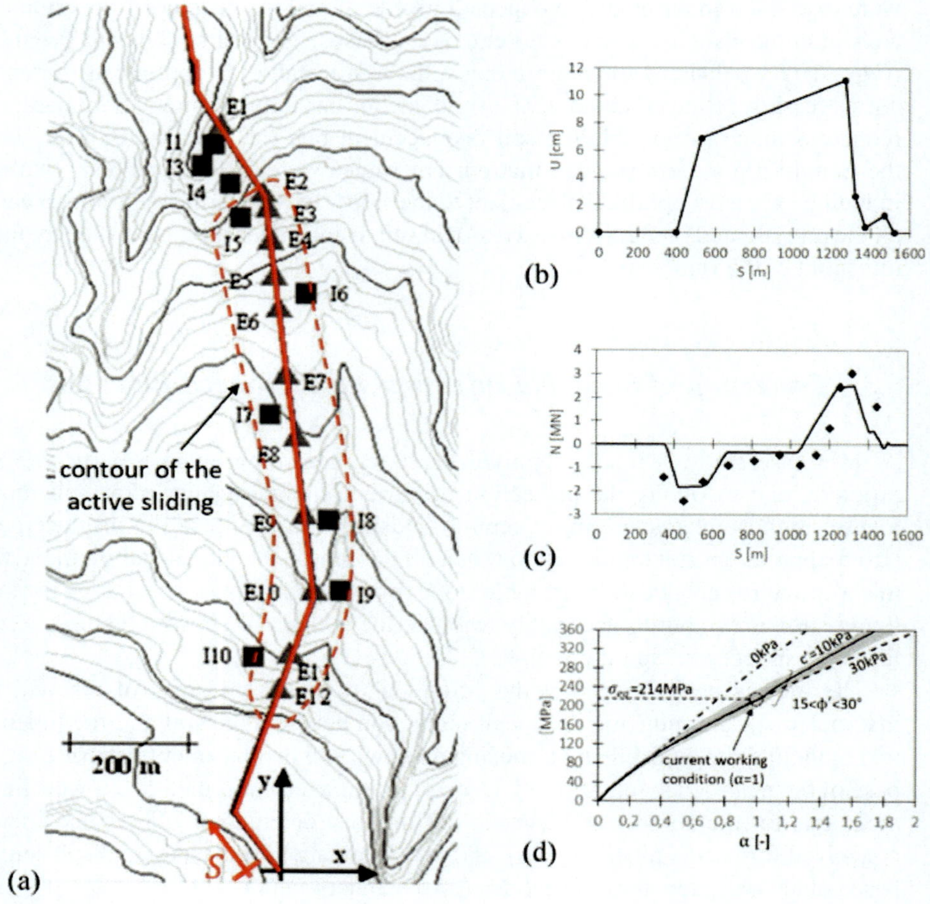

Fig. 10 **a** 3D pipeline layout; **b** measured pattern of superficial soil displacement along the pipeline; **c** computed values of the axial force N in the pipeline (full line) and comparison with the measured values (diamond); **d** parametric analysis on the future evolution of the average internal state of stress with the pipeline. Modified from [34]

component of soil displacement of about 8–10 cm (Fig. 10b), as it was measured by the inclinometric system (labels I1 to I10 in Fig. 10a) over a monitoring period of 6 years. Further details can be found in the aforementioned works [34, 44, 45]. The results of the analysis were validated by comparing the axial force arising within the pipe (representing the dominant action, for this considered case) with the values estimated by the extensometer readings installed along the pipe (labels E1 to I12 in Fig. 10a). A good agreement was observed between measured and computed data (Fig. 10c), with a peak of tensile axial force on the top of the hill (corresponding with a coordinate S approximately equal to 1300 m) and a concentration of compressive action at the base of the hill (S approximately equal to 500 m).

This preliminary analysis was extended by performing advanced parametrical analyse with respect to the adopted values of the soil mechanical parameters (which are often affected by high uncertainty) and to the amplitude of soil displacement profile (in order to foresee the future behavior of the system). These results are summarized in Fig. 10d in terms of the value of the maximum equivalent von Mises stress mobilized within the pipeline, in function of a scalar parameter a, scaling the amplitude of imposed soil displacement profile. Values of $a < 1$ reproduce past conditions of the system, $a = 1$ corresponds with the current (i.e. at the time of the analysis) state of the pipeline, and values of $a > 1$ represents future estimations. In particular, the effect of friction angle and cohesion values is investigated, showing that cohesion has here a larger influence on the value of the peak stress mobilized in the pipeline. This observation can be e.g. beneficial for optimizing more accurate site-specific soil characterization campaigns. Moreover, if a yielding value for the steel of 360 MPa is assumed, for the adopted set of parameters it can be observed that this pipeline can still accommodate an additional soil displacement of about 80% of the current value. In other words, the system can still tolerate 6–8 cm of additional soil displacement, which gives (by assuming a nearly constant average soil displacement rate over the years) a possible "safe" time window of about 4.8 years before structural failure. This information is obviously of fundamental importance for decision-making processes. Given the high computational efficiency of the numerical procedure, this value could also be continuously updated on the basis of the current monitoring readings, thus allowing to realize a real-time decision-making tool for this pipeline.

4 Conclusions

In the paper some real case studies of complex soil-structure interaction problems have been discussed in the light of sustainability concept. Some relevant considerations have been raised on the importance of adopting multidisciplinary approaches when addressing sustainability concepts, together with proper monitoring systems. Diffused reference has been done to the concept of "macroelement", consisting in upscaling the local behaviour of the soil REV to the macro scale of the structure, and representing then a very efficient computational tool. Once properly calibrated, macroelement approaches can reproduce real scale complex 3D SSI problems, thus allowing to perform even meaningful comparative analyses. These latter are particularly important in the preliminary phase of the design process, when strategic design choices can deeply increase the sustainability of the overall project. Given their high computational efficiency, they can even be employed for forecasting purposes or for real time monitoring, thus representing a valuable and time-adaptive decision-making tool. The results discussed in the paper, with a particular focus on the safety and efficiency of the constructions, proved the applicability of the proposed approaches to several domains, as the protection of historical heritage, of transportation infrastructure and of strategic lifelines, like e.g. buried pipelines affected by creeping landslides.

References

1. Brown RL (1981) Building a sustainable society. Norton, New York, W.W
2. Brundtland GH (1987) Our common future: report of the World Commission on environment and development. Oxford University Press, UK
3. Basu D, Misra A, Puppala AJ (2015) Sustainability and geotechnical engineering: Perspectives and review. Can Geotech J 52(1):96–113
4. Buchanan R (1992) Wicked problems in design thinking. Design Issues, VII I(2):1–21. https://doi.org/10.2307/1511637
5. Storesund R, Messe J, Kim Y (2008) Life cycle impacts for concrete retaining walls vs. bioengineered slopes. Proc Geo Congr 2008 Geotech Spec Publ No. 178 109(310):875–882. Doi:https://doi.org/10.1061/40971
6. Spaulding C, Masse F, La Brozzi J (2008) Ground improvement technologies for a sustainable world. Proc Geo Congr 2008 Geotech Spec Publ No. 178 111(310):891–898. Doi:https://doi.org/10.1061/40971
7. Holt DGA, Jefferson I, Braithwaite PA, Chapman DN (2009) Embedding sustainability into geotechnics. Part A: Methodology. Proc Inst Civ Eng 163(3):127–135. Doi:https://doi.org/10.1680/ensu.2010.163.3.127
8. Holt DGA (2011) Sustainable assessment for geotechnical projects. Ph.D. thesis, University of Birmingham, UK
9. Long JCS, Amadei B, Bardet J-P, Christian JT, Glaser SD, Goodings DJ, Kavazanjian E, Major DW, Mitchell JK et al (2009) Geological and geotechnical engineering in the new millennium: opportunities for research and technological innovation. Report of the Committee on Geological and Geotechnical Engineering in the New Millennium; Opportunities for Research and Technological Innovation, Committee on Geological and Geotechnical Engineering, National Research Council, The National Academic Press, Washington, D.C. http://www.nap.edu/catalog/11558.html
10. Pantelidou H, Nicholson D, Gaba A (2012) Sustainable geotechnics. Man Geotech Eng, vol 1. Institute of Civil Engineers, UK
11. Iai S (ed) (2011) Towards global sustainability. Springer, In Geotechnics and earthquake geotechnics towards global sustainability
12. Amorosi A (2020) The contribution of constitutive modelling to sustainable geotechnical engineering: examples and open issues. Rivista Italiana di Geotecnica 2:5–25
13. Montrasio L, Nova R (1988) Assestamenti di una fondazione modello sotto carico inclinato: risultati sperimentali e modellazione matematica. Riv Ital Di Geotec 22(1):35–49
14. Nova R, Montrasio L (1991) Settlements of shallow foundations on sand. Géotechnique 41(2):243–256. https://doi.org/10.1680/geot.1991.41.2.243

15. Butterfield R, Gottardi G (1994) A complete three-dimensional failure envelope for shallow footings on sand. Gètechnique 44(1):181–184
16. Butterfield R, Houlsby GT, Gottardi G (1997) Standardized sign conventions and notation for generally loaded foundations. Géotechnique 47(5):1051–1054
17. Gottardi G, Houlsby GT, Butterfield R (1999) Plastic response of circular footings on sand under general planar loading. Géotchnique 49(4):453–469
18. Cremer C, Pecker A, Davenne L (2001) Cyclic macro-element for soil-structure interaction: material and geometrical non-linearities. Int J Numer An-Alytical Methods Geomech 25(13):1257–1284
19. Cremer C, Pecker A, Davenne L (2002) Modelling of nonlinear dynamic behavior of a shal-low strip foundation with macro-element. J Earthquake Eng 06:175–211
20. Bienen B, Byrne BW, Houlsby GT, Cassidy MJ (2006) Investigating six-degree-of-freedom loading of shallow foundation on sand. Géotechnique 56(6):367–379
21. Gourvenec S (2007) Shape effects on the capacity of rectangular footings under general loading. Géotechnique 57(8):637–646
22. Gouvernec S (2007) Failure envelopes for offshore shallow foundations under general loading. Géotechnique 57(9):715–772
23. Grange S (2008) Modélisation simplifiée 3D de l'interaction sol-structure: application au génie parasismique. PhD Thesis, Institut National Polytechnique de Grenoble - INPG, 2008. Français
24. Grange S, Kotronis P, Mazars J (2008) A macro-element for a circular foundation to simu-late 3D soil-structure interaction. Int J Num Anal Meth Geomech 32:1205–1227
25. di Prisco C, Nova R, Sibilia A (2003) Shallow footing under cyclic loading: experimental behaviour and constitutive modelling. In: Maugeri M, Nova R (eds) Geotechnical analysis of the seismic vulnerability of historical monuments, Patron, Bologna, pp 99–122
26. di Prisco C, Massimino MR, Maugeri M, Nicolosi N, Nova R (2006) Cyclic numerical analyses of Noto Cathedral: soil-structure interaction modelling. Riv Ital Di Geo-Tec 2:49–64
27. di Prisco C, Vecchiotti M (2006) A rheological model for the description of boulder impacts on granular strata. Géotechnique 56(7):469–482
28. Paolucci R (1997) Simplified evaluation of earthquake-induced permanent displacements of shallow foundations. J Earthquake Eng 1(3):563–579
29. Paolucci R, Pecker A (1997) Seismic bearing capacity of shallow strip foundations on dry soils. Soils Found 37:95–105
30. Figini R (2010) Non-linear dynamic soil-structure interaction: application to seismic analysis and design of structures on shallow foundations. PhD Thesis, Department of Structural Engineering, Politecnico di Milano
31. Figini R, Paolucci R, Chatzigogos C (2012) A macro-element model for non-linear soil-shallow foundation-structure interaction under seismic loads: Theoretical development and experimental validation on large scale tests. Earthq Eng Struct Dy-Namics 41(3):475–493
32. Salciarini D, Tamagnini C (2009) A hypoplastic macroelement model for shallow founda-tions under monotonic and cyclic loads. Acta Geotech 4(3):163–176
33. Niemunis A, Herle I (1997) Hypoplastic model for cohesionless soils with elastic strain range. Mech Cohes Frict Mater 2:279–299
34. Galli A (2020) Macroelement approaches for geotechnical problems: a promising design framework? Riv Ital Di Geotec 2020(2):26–49
35. Magnani R (2014) La Missione segreta di Leonardo da Vinci, vol. 1 (in Italian) Asciano, Italy: Io sono edizioni
36. Galli A, Martinelli P (2016) Experimental characterization and numerical investigation on the Azzone Visconti bridge in Lecco (Italy). Procedia Eng 158:158–163
37. Martinelli P, Galli A, Barazzetti L, Colombo M, Felicetti R, Previtali M, Roncoroni F, Scola M, di Prisco M (2018) Bearing capacity assessment of a 14th century arch bridge in Lecco (Italy). Int J Arch Herit 12(2):237–256
38. Zani G, Martinelli P, Galli A, Gentile C, di Prisco M (2019) Seismic assessment of a 14th-century Stone Arch Bridge: role of soil-structure interaction. J Bridg Eng 24(7)

39. Zani G, Martinelli P, Galli A, di Prisco M (2020) Three-dimensional modelling of a multi-span masonry arch bridge: Influence of soil compressibility on the structural response under vertical static loads Engineering Structures 221:110998
40. Galli A, di Prisco C (2013) Displacement-based design procedure for slope-stabilizing piles. Can Geotech J 50(1):41–53
41. Galli A, Maiorano RMS, di Prisco C, Aversa S (2017) Design of slope-stabilizing piles: from ultimate limit state approaches to displacement based methods. Riv Ital Di Geotec 51(3):77–93
42. Kourkoulis R, Gelagoti F, Anastasopoulos I, Gazetas G (2012) Hybrid method for analysis and design of slope stabilizing piles. J Geotech Geoenvironmental Eng 138(1):1–14. https://doi.org/10.1061/(ASCE)GT.1943-5606.0000546
43. Galli A, Bassani A (2018) Innovative performance-based design of slope stabilizing piles for a railway embankment. Eur J Environ Civ Eng 22(1):99–121
44. Cocchetti G, di Prisco C, Galli A, Nova R (2009) Soil-pipeline interaction along unstable slopes: a coupled three-dimensional approach. Part 1: Theoretical formulation. Can Ge-Otechnical J 46(11):1289–1304
45. Cocchetti G, di Prisco C, Galli A (2009) Soil-pipeline interaction along unstable slopes: a coupled three-dimensional approach. Part 2: Numerical analyses. Can Ge-Otechnical J 46(11):1305–1321
46. Galli A (2005) Mechanical interaction between buried pipelined and landslides: small scale experimental analyses, numerical modelling and case studies. Ph.D. thesis, Department of Structural Engineering, Politecnico di Milano, Milan, Italy

Development of Flood Hazard Map of Upstream Johor River

Noor Farahain Muhammad Amin and Faridah Othman

Abstract Flooding has become a common occurrence in Malaysia, occurring every year in many states, particularly during the northeast monsoon. From 1926 until 2013, Johor State, in its most southern portion of Peninsular Malaysia, experienced severe floods. The Johor River watershed, on the other hand, was decimated by floods in December 2006 and January 2007. The floods flooded the relatively extensive catchment of the upstream Johor River, resulting in a substantial volume of discharge. The study's goal was to undertake river modelling and create a flood map for the Johor River upstream. The Johor River is 123 kms long and has a catchment area of 2,636 km. It starts from Mount Gemuruh and travels generally north–south before discharging into the Johor Strait. The data required in setting up this model includes the river spatial and geometrical data, hydraulics and hydrological data. The developing of the river model was starting by collecting data and insert the input data then the river model had been setup. The model had been calibrated and the results had been analyzed. The observed and simulated data have showed a reasonable agreement with the model. With a flood depth of 3.73 m and 100 ARI, Rantau Panjang is the most flooded area. The 100 ARI flood depth at Rantau Panjang is similar to the observed flood depth during the Johor River flood occurrences in 2007. The flood map River modelling can be a highly beneficial option because it is always possible to assess and anticipate with enough data.

Keywords River model · Flood diagram · Catchment area · Hydraulic · Upstream river

---

N. F. M. Amin (✉)
Faculty of Engineering & Built Environment, Mahsa University, Bandar Saujana Putra, 42610 Jenjarom, Selangor, Malaysia
e-mail: noorfarahain@mahsa.edu.my

F. Othman
Faculty of Engineering, University of Malaya, 50603 Kuala Lumpur, Malaysia

E. M. Nia et al. (eds.), *Sustainable Development Approaches*, Lecture Notes in Civil Engineering 243, https://doi.org/10.1007/978-3-030-99979-7_8

1 Introduction

Flooding is a natural hazard that has always happened in Malaysia. Practically every year in Malaysia, floods occur due to heavy rainfall. The technological innovation of today aggravate this risk. Flood behaviour is changing throughout the last 45 years as the severity and frequency of floods generated by human activities have increased [1]. Flooding affects about 29 000 km^2 (9%) of total land area and more than 4.82 million people (22%), with flood damage estimated at RM915 million per year [2]. Severe weather conditions and frequent floods are caused by the combination of extreme temperatures in equatorial regions, pressure gradients in these locations, and maritime exposure [3]. In Malaysia, there are 189 rivers that flow straight to the sea, 85 of which are prone to flooding [4].

During the floods of December 2006 and January 2007, the Northeast Monsoon, which occurs from December to March, caused significant flooding in various states across Peninsular Malaysia [5]. In four states, namely Johor, Negeri Sembilan, Melaka, and Pahang, these occurrences resulted in millions of dollars in losses and damages [6]. The first wave of floods occurred from December 19th to December 31st, 2006, while the second wave occurred from January 12th to January 17th, 2007. During the floods of December 2006 and January 2007, the Johor River watershed was the hardest hit. The floods flooded a huge area of the Johor River's catchment, which is relatively big.

The average monthly precipitation is 200 mm, however the 2006 storms had a 50-year return period, while the 2007 storms had a 100-year return period [3]. Kota Tinggi town, Kampung Baru Sungai Telor, Kampung Tembioh, Kampung Batu 25, Kampung Sungai Berangan, Kampung Semangar Dalam, Kampung Semangar Luar, Kampung Jawa, Kampung Kelantan, Kampung Sungai Sembilang, Kampung Sungai Sembilang, Kampung Sri Jaya, Kampung Rantau Panjang [5]. The goal of this work is to simulate the Johor River upstream and create a flood map. Johor is a state in Peninsular Malaysia that is located in the south. The Johor River Basin stretches from Tanjung Belungkor, in the state of Johor, to the slopes of Gunung Belumut, east of Kluang, and Bukit Gemuruh, in the north. This basin has a catchment area of around 2,690 km^2 and an upstream catchment area of about 1,130 km^2. The basin is mostly flat, with the exception of the northern and eastern regions, where terrain rises to a height of 500 to 600 mRL in the east and 1010 mRL in the north.

2 Methods

InfoWorks RS features comprehensive solution modelling of open channels, floodplains, embankments, and hydraulic structures as part of its hydrodynamic modelling programme. The rainfall-runoff simulation is also accessible in complete interactive views of data utilising both event-based and conceptual hydrological approaches, as well as geographical plan views, sectional views, long sections, spreadsheets, and time-varying graphical data. Any graphical or geographical view can access the underlying data. A sophisticated flood-interpolation model is overlaid onto an imported ground model to provide full flood mapping capability [7].

The Department of Irrigation and Drainage (DID) provided the cross-section profiles of the Johor River sub-basin. There are 21 river cross-section profiles, and the chainages interval used in this study was thousand metres, but smaller intervals down to 250 m were utilised in some areas that are more prone to flooding and more developed to compute the water level and flood map more precisely. Figure 1 depicts the position of 21 cross-sections along Rantau Panjang, which is located upstream of

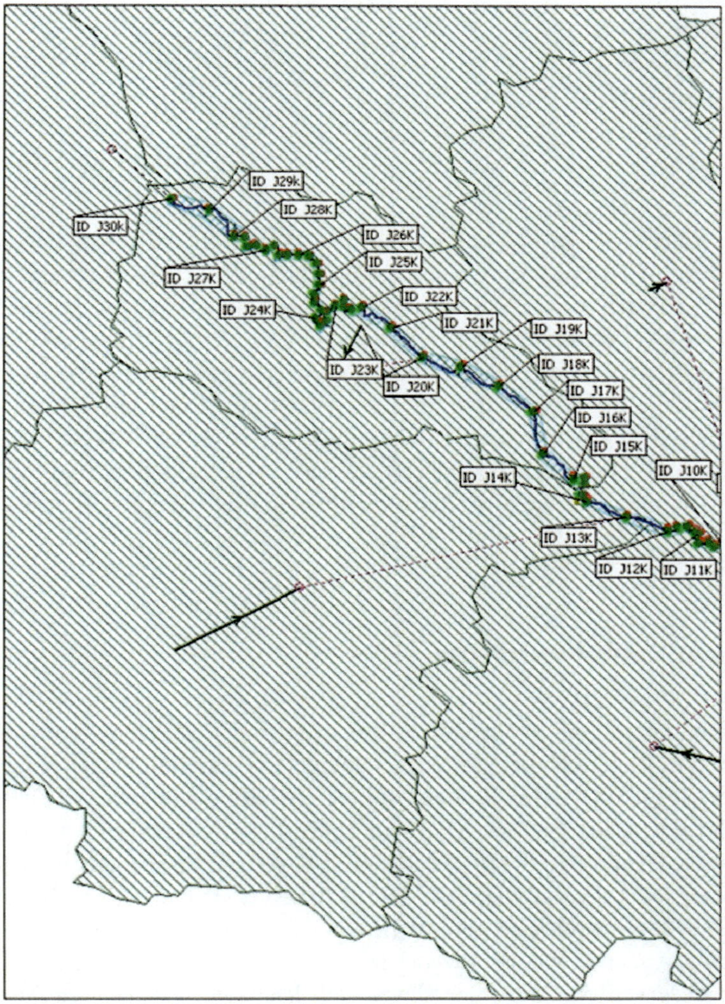

Fig. 1 The cross-sections inserted along upstream Johor River

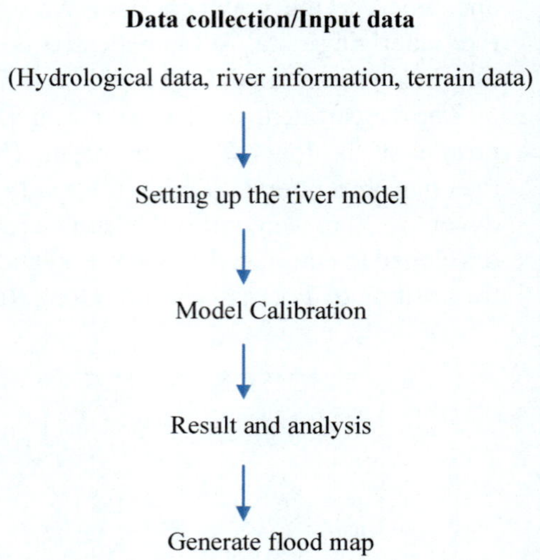

Fig. 2 Overall methodology in river modeling

Sungai Johor. The cross sections were created in AutoCAD format and converted to Shapefiles (SHP) data before being imported into the InfoWorks RS model database.

InfoWorks RS was used to import hydrological data such as streamflow, rainfall, and water level data. A model is built using these data and then simulated for calibration and verification. The input data at the upstream and downstream end points, as well as the tributaries along the upstream Johor River, are known as boundary conditions. The downstream boundary data was based on Seluyut confluence stage data, while the upstream boundary data was based on Johor River discharge data. Figure 2 summarises the procedures involved in modelling the river with InfoWorks RS.

3 Result and Discussion

The flooding region in the Johor River upstream stretches from around Bukit Gemuruh to Rantau Panjang. Figure 3 depicts the maximum and lowest water levels (blue lines) for 100 ARI in Rantau Panjang. The flood will occur when the maximum water level exceeds the river's left and right banks. In longitudinal section view, the high-water level along the upstream Johor River can be seen, as depicted in Fig. 4. The water level for the whole stretch of the river is depicted in this diagram. The simulated water level is on top, while the riverbed is on the bottom. Flooding will occur in any section when the water level is greater than the bank level.

The observed data for water level and discharge has been calibrated. Table 1 summarises the flood events with various ARI. The table shows that there is no flooding occurrence in Rantau Panjang for the return time of 2 ARI, but the flood depths for 50 and 100 ARI are relatively high, with 3.5 and 3.73 m, respectively.

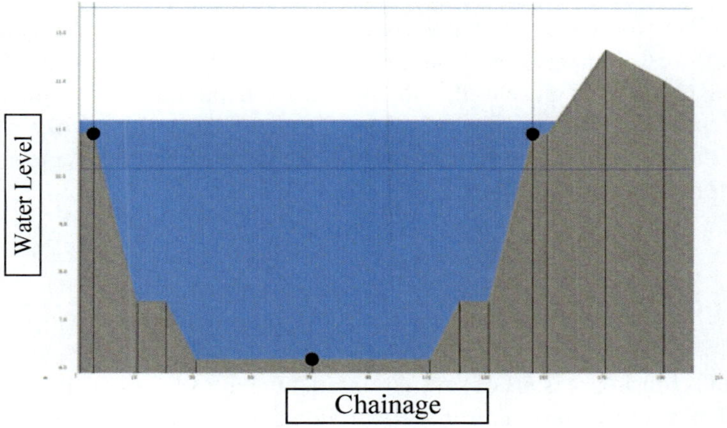

Fig. 3 Rantau Panjang cross section

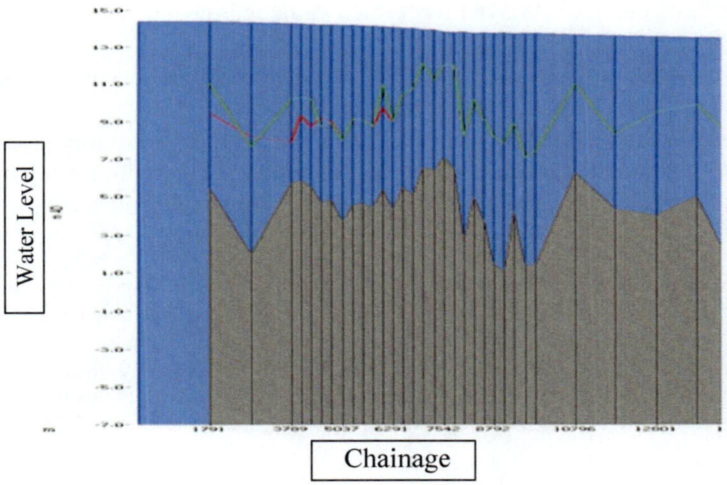

Fig. 4 Longitudinal section for Rantau Panjang

Table 1 Water level and flood depth for Rantau Panjang with different ARIs

ARI	Rantau Panjang		
	2	50	100
Max. River Level (M)	8.10	12.50	12.73
Danger Level (M)	9	9	9
Flood Depth (M)	No flooding	3.5	3.73

Rantau Panjang lies in the upper catchment region, with natural drainage, to serve the northwest drainage area from Semangar. The 100 ARI flood depth for Rantau Panjang (3.73 m) is close to the observed flood depth during the Johor River flood occurrences in 2007, which was 3.13 m. The flood danger map will make the region where the inundation area is determined more visible. Figure 5 depicts flood hazard maps for 2, 50, and 100 ARI.

4 Conclusions

- Finally, during seasons of severe rainfall, numerous locations in Rantau Panjang, which is upstream of the Johor River, flood several times a year.
- Rantau Panjang has the greatest flood depth of 3.73 with 100 ARI, making it the most flooded location.
- The flood depth of 100 ARI for Rantau Panjang is near to the observed flood depth for the 2007 flood events in Johor River.
- River modelling may be a highly beneficial option because assessment and prediction are always possible with enough data.
- Results showed that InfoWorks 1-Dimensional model is computational efficiency, ease of parameterization and easy representation of hydraulic structures in dealing with flows in large and complex networks of channels on the surface.

Acknowledgements The Department of Irrigation and Drainage Malaysia and the Regional Humid Tropics and Water Resources Centre for Southeast and the Pacific collaborated on this study (HTC). The authors would like to express their gratitude to Mahsa University for hosting the conference, the University Malaya Research Grant (UMRG), [RP017C-15SUS], and the Minister of Higher Education's Trans-disciplinary Grant Scheme (TRGS), [TR001B-2015], for financial support, and everyone involved for their guidance and support during the research period.

References

1. Elfithri R, Halimshah S, Abdullah MP, Mokhtar M, Toriman ME, Embi AF, Ramzan NM (2017) Pahang flood disaster: the potential flood drivers. Malays J Geosci 1(1):34–37
2. Mohd ET, Norbaya H, Mohd KAK, Abdul JH, Muhammad BG, Asyaari M, Nor AAA (2015) Assessment of water salinity model using hydrodynamic numerical modelling in estuary of Selangor River, Malaysia. Malays J Anal Sci 19(5):1109–1119
3. Abdullah J, Muhammad NS, Julien PY, Ariffin J, Shafie A (2018) Flood flow simulations and return period calculation for the Kota Tinggi watershed, Malaysia. J Flood Risk Manag 11:S766–S782
4. Zakaria NA, Ghani AA, Chang CK (2014) MSMA 2nd edition–application of green infrastructures for solving sustainable urban stormwater management challenges. Diakses daripada https://www.water.gov.my (Tarikh akses 12 Januari 2017)
5. Wahab HA, Yaacob N (2016) Pengurusan Sungai Mengikut Undang-Undang (UUM Press). UUM Press

6. Othman, Faridah, Amin, Muhammad, Farahain, Noor, Mi Fung, Lau, Elamin, Mohamed, Eldin, Alaa (2013) Utilizing GIS and infoworks RS in modelling the flooding events for a tropical river basin. Paper presented at the Applied Mechanics and Materials
7. Jajarmizadeh M, Harun S, Salarpour M (2012) A review on theoretical consideration and types of models in hydrology. J Environ Sci Technol 5(5):249–261

Smart City Technology in the Fisherman Tourism Village in the Colorful Village, Jakarta, Indonesia

Boy Subirosa Sabarguna

Abstract The existing development of tourism villages related to tourism development are: an overview for situation analysis, such as: Head of Cilincing Village that the change in Fisherman's Village from what was slum and dirty, will have positive impacts including the potential of the tourism sector to become a tourist attraction, it needs to be continued until international level. This study investigated the development of coastal tourism villages as a reference to improve the welfare of the region, as the main priority of the central government of Indonesia. The development includes the development of an International Class Fisherman Tourism Village based on Smart City Technology, which is supported by: Clean Coastal Environmental Health and Waste Recycling, Creative Economy, Art, Culture and Craft Studios, and Organizational, Management and System Information Strengthening. It was also fostered by Marine Medicine Study Program Development Team, Department of Community Medicine, Faculty of Medicine, University of Indonesia, Readiness of the University of Indonesia Directorate of Community Service and Development, who can oversee the program and the readiness of Village. The study area was located at Colorful Village Fisherman's Village RT 012/RW 04 Cilincing Village, Cilincing District, North Jakarta, Jakarta Province, Indonesia. The study proposes a development process in the context of realizing a fishing village based on Smart City Technology, for tourism and the welfare of coastal communities.

Keywords Smart city technology · Fishing village · Tourism · Coastal areas

1 Introduction

The existing development of tourism villages related to tourism development are: an overview for situation analysis, such as: Head of Cilincing Village that the change in Fisherman's Village from what was slum and dirty, will have positive impacts including the potential of the tourism sector to become a tourist attraction, it needs to

B. S. Sabarguna (✉)
Department of Community Medicine, Faculty of Medicine, Universitas Indonesia, Jakarta, Indonesia
e-mail: sabarguna08@ui.ac.id

E. M. Nia et al. (eds.), *Sustainable Development Approaches*, Lecture Notes in Civil Engineering 243, https://doi.org/10.1007/978-3-030-99979-7_9

be continued until international level. This is related to the development of community organizations, which has been running since the Village Community Health Education (PKMD) [1] since 1970 and continues to grow slowly. In 1984, in line with the Family Planning program, a pattern of active community participation was developed starting with the establishment of the Posyandu in 1985 which continued until 1990. Now is the time to be more active in developing the Posyandu for the Elderly, whose number is increasing and they need to remain productive in the family and community.

Community participation has not been developed, it is still low, so there is a lack of readiness to be independent and lack a sense of belonging to the renewal program. In Smart city conditions, seniors can be productive elderly using productivity for elderly monitoring [2] so that productivity can be measured easily without a burden. The application from the beginning uses the smart city concept to make it clear and activities that meet the future [3]. Tourism village development [4] is related to tourism development as an illustration for situation analysis. The Village Head of Cilincing, Sugiman said that the change in the Fisherman's Village from what was previously slum and dirty, will have a positive impact, including the potential of the tourism sector to become a tourist attraction. "With this colorful and neat, it is hoped that visitors can stop by. There is an attraction," said Sugiman, Wednesday (4/24/2019). Cilincing Fisherman Village painted in colorful colors like a rainbow (Fig. 1), attracting tourists to visit Wednesday, 24 April 2019 17:44-warta kota-Nur Ichsan [5]. Regarding BPJS [6], it is recommended that the government map the informal sector fishermen, the fishermen need to be considered as PBI participants, socialization and JKN media need to be carried out, and BPJS Health picks up the ball in terms of JKN registration [7].

Fig. 1 Colorful Village at Cilincing, Jakarta

The research shows that: (1) Fisherman groups 1 and 2 already know about climate change, so they can anticipate what actions must be taken to deal with climate change that occurs, (2) Fisherman groups 1 and 2 already know the function of cooperatives as an alternative and way to solve difficulties in meeting their daily needs and to increase their business capital. In terms of history, [8] the Cilincing area is located east of Tanjung Priok Harbor, today it is a sub-district. Cilincing sub-district, including the North Jakarta Municipality area. The name Cilincing is taken from the name of a tributary that flows from south to north, dividing the area.

2 Methods

From the problems above, efforts are made to develop the community to continue what already exists to become an international class fishing tourism village, because many foreign tourists will cross to the thousand islands. Thus the development is directed, at the following. (1) Development of tourism villages, is a reference in order to improve regional welfare, as the main priority of the central government, now it needs to be strengthened based on Smart city [9]. (2) The establishment of the Elderly Posyandu is supported by integrated programs, including Tourism Villages as the development of Creative Economy and Art Studios, Garbage Banks, which simultaneously run will have a more real impact on the welfare of families who have the elderly and the welfare of the community in the village. (3) Environmental conditions are clean, people are aware of health and the need for BPS, and the understanding of cooperatives already exists, so it is necessary to continue to realize it, not just hope. (4) Social Engineering training which capacity building in assisting in preparing for community participation and generating a sense of belonging to the assistance provided. (5) The establishment of tourist villages is important today, related to tourism and empowering the creative economy as well as strengthening the Indonesian economy. (6) Strengthening the organization, management and information system so that the fishing village communities are able to develop independently, and can ensure the sustainability of the program.

The success of various integrated programs requires careful preparation so that the stunting alleviation program can run continuously. Knowledge-based activities, the application of technology, are needed to support clear and traceable progress. In addition, continuous efforts are needed, including using a Social Engineering Model that can be applied and used as an example in other regions or regions.

From the description above, it can be abbreviated, as follows: (1) The village is clean and organized, it needs to be fostered and continued by guiding the sustainability of the productive elderly program so that it is integrated and clearly implements Smart city in real terms; (2) There are already seeds of fishermen's economy, health and BPJS as well as cooperatives, all that remains is to strengthen the organization, management and information system so that the fishing village community is able to develop independently, and can guarantee the sustainability of the program; (3) To build an International class Tourism Village, which will support the Thousand Island tourism, just need to develop it further with a social engineering creative economy program.

Objective related to the following matters: (1) The development of coastal tourism villages is a reference in order to improve the welfare of the region, as the main priority of the central government; (2) Environmental conditions are clean, people are aware of health and aware of elderly productivity; (3) Organized Smart City Technology which is capacity building in mentoring to prepare participation that evokes a sense of belonging to the assistance provided and the establishment of international class tourist villages, especially related to tourism and creative economic empowerment, as well as strengthening the Indonesian economy in the post-Covid-19 pandemic.

3 Results and Discussion

In general as follows and includes: (1) The Economy aspect in health and tourism, (2) Art, Culture and Craft Studios, and (3) Organizational, Management and Information System Strengthening development of an International Class Fisherman Tourism Village based on Smart City Technology [10], which is supported by: (4) Coastal Environmental Health [11] and Waste Recycling, (5) Creative Thinking and Smart city implementation. The following description in detail as below, Current conditions are: (1) The village is clean and organized, it needs to be fostered and continued by guiding sustainability; (2) There are already seeds of fishermen's economy, health and BPJS as well as cooperatives, just need to strengthen them; (3) Build an International Class Tourism Village, which will support Thousand Island tourism, just need to develop it further. There are 5 important things related to Smart city, as a marker as one clear evidence, in accordance with the expectations of a good environment in Indonesia, namely: [12] (1) Using the framework of ICT (Information and Community Technology); (2) To develop; (3) To apply; (4) To encourage the continuity of implementation; (5) To solve urbanization challenges. From the description above, it can be abbreviated, as follows: (1) The village is clean and organized, it needs to be fostered and continued by guiding the sustainability of the productive elderly program so that it is integrated and clearly implements Smart city in real terms. [13] (2) There are already seeds of fishermen's economy, health and BPJS as well as cooperatives [14], all that remains is to strengthen the organization, management and information system so that the fishing village community is able to develop independently, and can guarantee the sustainability of the program; (3) To build an International class Tourism Village, which will support the Thousand Island tourism, just need to develop it further with a social engineering creative economy program [15].

4 Conclusion

This study pinpoints the importance and opportunities of the the fishermen village that was actually situated in slum environment, to develop into tourism a tourist attraction in an international level. To implement that, this study highlights the urgent of elevating the development of coastal tourism villages to improve the welfare of the area by the central government. The coastal tourism seen as potentials to attract the creative economic empowerment and strengthening the Indonesian macro economy in the post-Covid-19 pandemic generally. This study suggests the development of an International Class Fisherman Tourism Village based on Smart City Technology, which the implementation could be supported by: the Clean Coastal Environmental Health and Waste Recycling, Creative Economy, Art, Culture and Craft Studios, and Organizational, Management and System Information Strengthening. The smart city concept in the fishermen village could be an alternative strategic solution to solve the local and macro socio economic society problems in facing the current pandemic challenges.

References

1. Sabarguna BS (2015) Posyandu, Pembangunan Masyarakat Desa. Jakarta: Sagung Seto
2. Pang S-Y, Arivalagan Y (2020) These countries are most ready to deal with ageing populations. https://www.weforum.org/agenda/2020/02/what-are-japan-and-singapore-doing-about-ageing-population/ (2021-09-19, 7:55 PM)
3. Thales (2021) Secure, sustainable smart cities and the IoT. https://www.thalesgroup.com/en/markets/digital-identity-and-security/iot/inspired/smart-cities
4. Anna Suci P (2020) Kemenpanrekraf bakal minta kepada desa di Indonesia bikin Desa Wisata. https://nasional.kontan.co.id/news/kemenparekraf-bakal-minta-kepada-desa-di-indonesia-bikin-desa-wisata (2021-1-25, 12:31 PM)
5. Junianto H (2019) Kampung Nelayan Cilincing Dicat Warna Warni Seperti Pelangi, Menggugah Minat Wisatawan Berkunjung. https://wartakota.tribunnews.com/2019/04/24/kampung-nelayan-cilincing-dicat-warna-warni-seperti-pelangi-menggugah-minat-wisatawan-berkunjung
6. Dara L, Suryawati C, Fatmasari EY (2016, Januari) Gambaran Sektor Informal Nelayan Keluarahan Cilincing Jakarta Utara Sebagai Calon Peserta BPJS Kesehatan. Jurnal Kesehatan Masyarakat 4(1):2356–3346
7. https://e-journal.jurwidyakop3.com/index.php/jurnal-ilmiah/article/view/251/219 (2020-1-2, 10.00 PM)
8. Statistik BP (2020) Kota Jakarta Utara Dalam Angka 2020. Badan Pusat Statistik Indonesia
9. Winkowska J, Szpilko D, Pejić S (2019) Smart city concept in the light of the literature review. Eng Manag Prod Serv 11(2):2019
10. BINUS University Malang (2021) Penerapan Smart City Di Indonesia. https://binus.ac.id/malang/2021/04/penerapan-smart-city-di-indonesia/ (2021-09-19, 8:32 PM)
11. EU SCIENCE HUB. https://ec.europa.eu/jrc/en/research-topic/coastal-and-marine-environment (2021-09-19, 8:53 PM)
12. Ministry of Environment and Forestry, Republic of Indonesia (2020) *The State of Indonesia's Environment 2020.* Ministry of Environment and Forestry, Republic of Indonesia

13. Rosa D, Sabarguna BS, Abdurakhman, Dewi DK, Pusposari D, Maryani D, Nursasi AY, Christia M, Daryanti S (2020) An integrated toddler nutrition strengthening system in an area based on the computerization system and social engineering development (Suatu Sistem Penguatan Gizi Balita Terpadu Pada Suatu Daerah Berbasis Sistem Koputerisasi dan Pengembangan Social Engineering). Universitas Indonesia
14. Mauricio A, Barrero F (2021) Identification of synergies and conflicts in coastal and marine uses in Colombian Pacific: a Spatial Multi-Criteria Analysis. Doi: https://doi.org/10.26359/costas
15. European Union (2021) The EU Blue Economy Report 2021. Directorate-general for maritime affairs and fisheries

Hospital Ship, Hospital Clinic and Telemedicine for Disaster Medicine Adaptation at Coastal Region

Boy Subirosa Sabarguna

Abstract Many coastal areas in Indonesia are spread widely and in various groups, which require special preparation and handling in times of disaster related to medical services. With the dean, appropriate adaptations are required for planning, preparation, implementation and post-disaster. This study investigates the medical service system in the form of hospital ships and hospital clinic in boats that are ready to operate in the state of preparation, implementation and post-disaster with the support of telemedicine. Thus, medical services can be used during normal and disaster situations. The study used the materials related to the arrangement of hospital ships and boat clinics that meet optimal standards that are ready for normal and disaster use. Medical equipment, medical support, drugs and operational procedures for both normal and disaster situations. Doctors and other officers in accordance with the required profession, with special training for disasters. The result shows an integrated information system and telemedicine system that uses voice recognition in times of disaster, supported by trained equipment and personnel. The method used is in the form of needs assessment and feasibility studies as well as making proposals for funding, thus trying to attract the participation of related parties, especially BAPPENAS, the Central Government, Regional Governments and BNPB. This study proposes suggestion with the support of a needs assessment and a feasibility study, which fulfills the readiness of facilities, doctors and other health workers who are ready for normal times and during disasters. Thus, it is ready to be discussed and ready to be worked on, because coastal conditions have begun to sink due to extreme climate change today. The recommendation is suggested to be discussed with relevant stakeholders will encourage awareness of the importance of Hospital Ship, Hospital Clinic and Telemedicine for Disaster Medicine Adaptation in Coastal Region, so that climate change that changes the current coastal area in terms of medical services can be prepared.

Keywords Hospital ship · Hospital clinic · Telemedicine · Disaster medicine adaptation · Coastal region

---

B. S. Sabarguna (✉)
Development Team for Marine Medicine Specialist-1 Study Program, Department of Community Medicine, Faculty of Medicine, Universitas Indonesia, Kota Depok, Indonesia
e-mail: sabarguna08@ui.ac.id

E. M. Nia et al. (eds.), *Sustainable Development Approaches*, Lecture Notes in Civil Engineering 243, https://doi.org/10.1007/978-3-030-99979-7_10

1 Introduction

As background to the making of Telemedicine for Mobile Hospital During Disaster, such as the following: (1) Disaster, the existence of Mobile hospital complete with Telemedicine and Clinical Decision Support System, will accelerate disaster management effort related to medical service; (2) Health care facilities will be moving easily implemented, it will be able to immediately service the movement of Mobile Hospital and supported by network of Infrared and Communication technology with telemedicine and Clinical Decision Support System; (3) The health care provider will difficult and less, with quality services and the equivalent of face-to-face; (4) The competence of personnel will be limited, the use of telemedicine will help fill this needs; in this case the implementation will be carried out: training, technical guidance, monitoring and evaluation.

The eligible subject includes the followings; (1) Innovation, related to the implementation of medical disaster management, which is needed immediately, quickly and qualified. Related to Clinical Decision Support System innovation implemented in the form of Mobile Hospital on disaster occurrence. The application of innovation is also related to the application of telemedicine in a concrete form that will be one way to improve health, especially in the area that is facing disaster. (2) Impact and outreach, very important for Indonesia and Asia Pacific associated with the Ring of Fire is easy to disaster, and the existence of a large human movement during Eid al-Fitr. The immediate impact will be on people facing disaster with medical treatment, using Mobile Hospital, Clinical Decision Support System, and Telemedicine. Reaches very broadly in large areas of Indonesia with many islands and great potential disasters as they are in the Ring of Fire, and in populations of large human movements during Eid al-Fitr. (3) Mobile hospital and its equipment will be easily replicated, because the system and components are ready. Related systems and equipment Mobile Hospital, tools, Decision Support System and Telemedicine, will be easy to apply when this product has occurred due to the technology and the existing and prepared component. In Indonesia there are at least 5 major islands such as Sumatra, Java, Kalimantan, Sulawesi and Papua, so need 25 pieces. If every provincial minimum, need 1 set then it takes 35 pieces. (4) Sustainability in Indonesia is ready because there are PNBP and BPBD which is an institution, which has staff and budget. In Indonesia, there are BNPB (National Guard on Disaster) and 35 BPBD (State Guard on Disaster), then operational execution already existing organization and budget, live in linked and socialized and trained to use effectively and efficiently. (5) Business Plan and Market Opportunity, for Indonesia it takes ± 100 pieces and In Asia pacific related Ring of Fire ± 1000 required.

Based on the above background, the objective of Telemedicine and Clinical Decision Support System for Mobile Hospitals on disasters such as the following. (1) Make Mobile Hospital with the optimum standards so easily moved or transferred at the current location of the disaster area, which support with telemedicine and Clinical Decision Support System. As Hospital Ship, Hospital Clinic or Clinical Boat as means of transportation for medical and health service tools, materials and personnel in costal and marine areas. (2) Provide support telemedicine network telemedicine network are: RSUI DEPOK (Depok Universitas Indonesia Hospital), BNPB (National Guard for Disaster Management) and BPBD (Local Guard for Disaster Management). (3) Create a pattern of service (Clinical Decision Support System Network) to do a wider range of competence and of better quality with Mobile Hospital, Hospital Supervisor and Controller. (4) To make System for Integrated Telemedicine Network [1], national, local and hospitals, the Hospital Mobile, Hospital Supervisor and controller are: RSUI DEPOK (Depok Universitas Indonesia Hospital) and BNPB (National Guard Disaster Management). (5) To make Integrated in Operational Services System using Clinical Decision Support System for getting efficiency and effectivity.

2 Methods

The concept is described as follows: (1) Mobile Hospital, a hospital with the most important services, with minimal appliance standards, which can move or be moved, supported by telemedicine, the number 1 or more; (2) Telemedicine Network, is a network that is the liaison with the use of telemedicine, so that the unity of operations, both locally and nationally, (3) Hospital Supervisor, the Hospital of class C or class B relative near to the disaster site, which is where referral and include telemedicine network. Hospital supervisor can oversee Engaged 1–10 Hospital; (4) controller, a general national control of BNPB (National Agency for Disaster alleviation) and the hospitals nationwide by the RSCM (Cipto Mangunkusumo Hospital) and RSUI (Universitas Indonesia Hospital); (5) Using Clinical Decision Support System for Operational Services to get effectivity and efficiency. Theory will be related to: (1) Medical Disaster Information System, (2) Medical Information System during Disaster, (3) Telemedicine Network [2], (4) Mobile Hospital, (5) Hospital Supervisor, (6) Control, (7) Integrated Care System, in proportionally supporting the implementation of the service. Telemedicine and Decision Support System, can be seen as Fig. 1 below.

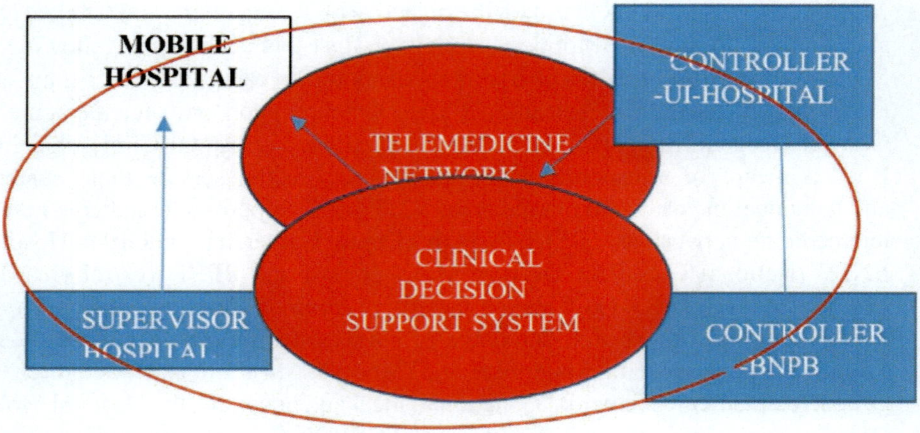

Fig. 1 Telemedicine and decision support system figure of telemedicine and mobile hospital, can be seen at: [3, 4]

Appropriate with focus areas, will be associated with the Good Health related to disaster, the overall interconnected between: (1) telemedicine, (2) Clinical Decision Support System, (3) mobile hospital, (4) disaster that effect to medical problems [5]. Telemedicine, a science related to computer networks, communications and information technology, which can assist in the implementation of medical services in remote conditions; with will be very helpful in order to service existing barriers within. Telemedicine, will be very useful in order to support medical aid during disaster that are limited in terms of distance, included in this trouble competence of experts, especially doctors. Clinical Decision Support System, is the implementation of research results that have 4 patents (2 Granted + 2 Registration) and 14 copyright (3 for Disaster, 3 for ECG, EEG and USG, 8 for Hospital Services) provide direction for the manufacture and operation of Mobile Hospital fast, easy and good quality. Mobile Hospital is a hospital that can move or be moved at certain locations that need because there are communication difficulties and the location of the land and sea.

Thus the use of a vehicle which can be moved it would be easy to service at the time of the disaster, which destroyed communication lines, to be appointed by the truck, and so on by helicopter so the hospital will be ready to use. At the time of the disaster who need medical services, the hospital facilities and communication network not enough, it needs the health personnel especially doctors, telemedicine systems will help: bring experts for consultation, monitoring and through mentoring medical treatment from the hospital supervisor to be a part of hospital delivery system: (1) Clinical Decision Support system; is a software for medical services in hospitals; (2) Telemedicine; will be combined as a network of information systems and service networks in disaster situations that address the medical and health problems of affected people; (3) It will be a mobile hospital that can be used with, fast, easy and quality.

3 Result

Border impacts, which will arise in the field: Medical Services during a disaster, medical information system and telemedicine, Mobile Hospital for times of disaster, such as the following.

1) Medical Services on During Disaster, require specification specifically: (a) more on emergency services, (b) with limited facilities and personnel, (c) requires a high speed, thus will provide fast service, and meet the minimum quality:
2) Systems Medical Information, such as medical records, medical consultation, up to supervision and mentoring therapy is often a big problem, because of the limitations of the communication, the use of telemedicine will be a solution that provides support for a fast, easy and can be done according to the existing conditions:
3) Mobile Hospital, is required because of limited transport, and difficulties in the field, then the model of the mobile hospital would be a practical way out and ready to use at any time, so as to facilitate the planning and implementation:
4) Lack of competence in the case of referral, to service in the field will be difficult to obtain physical, takes time and effort that is difficult, then their telemedicine [6], for: consultation, monitoring through mentoring therapy can be done, as a general surgeon can be mentored in surgery orthopedic difficult, with the presence of doctors in the orthopedic Hospital Supervisor.

In a wider scale will give more directives: (a) Planning a clearer, services, facilities and personnel of medical services during a disaster, (b) Provide the reference guide that can be prepared Standard Operating Procedures use of telemedicine in the mobile hospital, (c) Provide direction for staff training who are ready to work on the use of telemedicine in the mobile hospital, (d) Develop coordination and control patterns between BNPB and BPBDs, Hospital Controller with Supervisor Hospital. Referrals will be a floating models that will be useful for Indonesia and other countries in the world, because of the disaster that requires medical services is currently a lot going on.

4 Discussion

4.1 Sustainability

Sustainability will be done in two types as follows: (1) Decision Support System and Telemedicine Network and Mobile Hospital and medical services during disaster as a model, will be submitted to the BNPB [7], so that will be used for services. Otherwise it will become a model to be made more numerous; with will provide continuity of development with a more extensive, sophisticated and appropriate to the needs of the disaster area. (2) Models created can now be used as a training center for: (a) Personnel associated with medical services during disaster, (b) Training of personnel working on the mobile hospital, (c) Personnel who operate telemedicine,

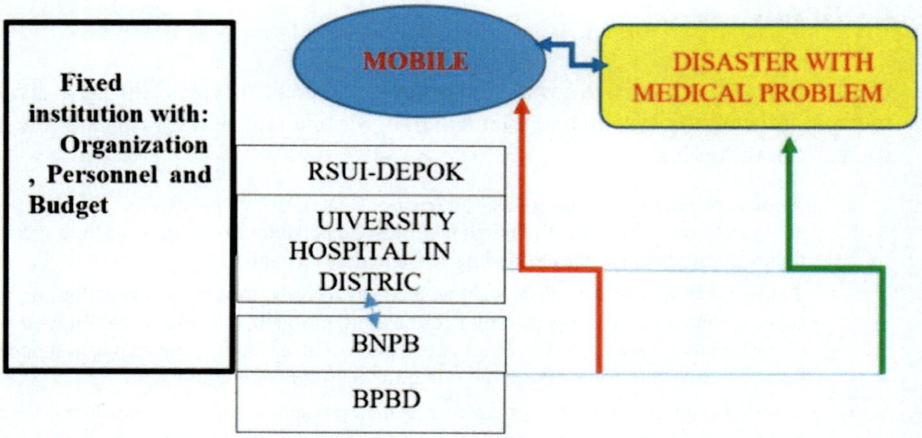

Fig. 2 Sustainability program

(d) Controller in BNPB and BPBDs, RSUI [8] and other Supervisor Hospital. Both of the above will be the center of national development and training ready to develop an international scale (Fig. 2).

4.2 *Innovative Approached*

Innovative approached, held on four main topics, namely: (1) Making telemedicine system during disaster, is the development of telemedicine that are tailored to the specific needs during disasters, a new application made by design and scientific stage, (2) Clinical Decision Support System, to support service management and resource management for operational daily, (3) Preparation for a mobile hospital during a disaster, a scientific approached which is action research, to create a mobile hospital that is suitable for the time of the disaster who need medical services, will generate the appropriate model, associated layout, appliance and matching telemedicine network, (4) Networking System [9] of controller in Telemedicine between PNBP and BPBDs, RSCM with Supervisor Hospital, which is a scientific study and applied that blends theory with application, resulting in effective and efficient systems, which will be new to the system of control of medical services at the time of disaster require medical care [10, 11].

5 Conclusion

This study highlights the significance of medical equipment, medical support, drugs and operational procedures for both normal and disaster situations. It also suggests doctors and other officers in accordance with the required profession, with special training for disasters, an integrated information system and telemedicine system using voice recognition in times of disaster, supported by trained equipment and personnel. The study recommends a proposal with some plan framework to be discussed with relevant stakeholders will encourage awareness of the importance of Hospital Ship, Hospital Clinic and Telemedicine for Disaster Medicine Adaptation in Coastal Region, so that climate change that changes the current coastal area in terms of medical services can be prepared. This finding could be used for scenario of strategizing disaster medicine adaptation at coastal region especially in Indonesia context.

References

1. Sabarguna BS (2012) E-mini hospital for implementation for mass and cooperation. IJDMS
2. Sabarguna BS (2016) Sistem Informasi pada Peraltan Medis di Rumah Sakit. UI Press, Jakarta
3. Saravanan S, Anbu Rajan M, Venkatraman A, Sriraam N, and Thirusakthi Murugan P (2011, July–December) Telemedicine using computer communication network. Int J Comput Sci Commun 2(2):623–631
4. Jarek B (2016) A mobile hospital—Its advantages and functional limitations. Int J Saf Secur Eng 6(4). https://doi.org/10.2495/SAFE-V6-N4-746-754
5. Sabarguna BS (2015) Sistem Infromasi Medis untuk Mitigasi Bencana (Medical Information System for Disaster Mitigation). No. IDP0050757
6. Sabarguna BS (2015) Sistem informasi Medis untuk Saat Bencana (Medical Information System During Disaster). C00201501403
7. Rencana Strategis BNPB Tahun 2020–2024
8. Deddy S (2015) Jepang Sumbangkan Sistem Medis Canggih untuk RSCM. Available at: http://www.cnnindonesia.com/teknologi/20150528070816-199-56186/jepang-sumbangkan-sistem-medis-canggih-untuk-rscm/ (July 27, 2016, 15.17)
9. Sabarguna BS (2011) Networking system for hospital (English). https://nulisbuku.com
10. Sabarguna BS (2015) Usage of "Flowchart of Diagnosis and Treatment" software in medical education. ICHMI, Zuruch, Switzerland, 2015, to support Clicical Information System
11. Health Disaster Management (1999) Methods used for disaster medical research, health disaster management guidelines for evaluation and research in the Utstein style. Available at: https://wadem.org/wp-content/uploads/2016/03/chapter_2.pdf (July 7, 2016, 06.18)

The Concept of Sustainability in the Formation of Unplanned Settlement Spaces in Depok Lama, West Java

Rakhmanita, Edi Purwanto, R. Siti Rukayah, and Arief Rahman

Abstract As a village that later developed into an area that resembled a colonial city at that time, Depok Lama had a concept of residential space that was different from the colonial settlements that existed during the Dutch colonial period. In other colonial settlements, the concept of settlement was the result of a design from the Dutch government, in the Depok Lama settlement the concept of residential space was formed from the activities of citizens directed by a will. This method is used to find a detailed picture based on facts about the beginning of the formation of Depok lama settlement through collecting Depok Lama maps, Chastelein's will, and observing city artifacts. The results of this study note that the concept of space that was applied at the beginning of the formation of the Depok Lama settlement was an unplanned concept of sustainability, prioritizing easy access to various main facilities in the Depok Lama settlement, forming a comfortable environment for the residents of Depok Lama in building their lives and improving they socio-economic quality.

1 Introduction

By 2092, the world population concentrated in urban areas is projected to reach 100% [1]. In 1898, long before the United Nations and several international organizations began discussions on the topic of sustainability and sustainable development, Ebenizard Howard had introduced the topic of sustainability in the concept of the Garden City.

In Indonesia, the concept of the Garden City was brought by the colonialists and applied in the planning of cities with the help of urban planners and architects, to

Rakhmanita (✉)
Doctoral Program in Architecture and Urban Sciences, Diponegoro University, Semarang, Indonesia
e-mail: rakhmanita.ug@gmail.com

E. Purwanto · R. S. Rukayah
Department of Architecture, Faculty of Engineering, Diponegoro University, Semarang, Indonesia

Rakhmanita · A. Rahman
Department of Architecture, Faculty of Civil Engineering and Planning, Gunadarma University, Depok City, Indonesia

E. M. Nia et al. (eds.), *Sustainable Development Approaches*, Lecture Notes in Civil Engineering 243, https://doi.org/10.1007/978-3-030-99979-7_11

create structured and well-planned cities. A city with a Garden City concept has complete facilities with an affordable distance on foot so that people are comfortable doing activities [2, 3].

Interestingly, Indonesia has a village where plantation slaves lived, which was established by colonial rulers who seemed to have a Garden City concept, but not through structured planning like other colonial cities in Indonesia. Currently, the village was part of the city on the outskirts of the capital. This area is known as Depok Lama area in Depok City, West Java. This area has a history of 307 years and is still in the midst of the rapid development of Depok City.

Since it is assumed that there is no structured plan for the formation of the space, this research paper describes how the people there form the space, and create the existing facilities. The purpose of this study was to determine the initial concept of the development of the Depok Lama residential area with the concept of a garden city which was declared a sustainable city.

In conducting this research, we used a qualitative description method based on a historical approach. This is done by examining built and undeveloped spaces from 1917 to 2010, starting with activities in the area, as well as reading and analyzing what Cornelis Chasteline left as the founder. The will may be used as a source of research [4], to help to understand what Cornelis Chastelein wanted in the lives of slave workers. The last stage is observing the field and collecting old and new photos of the city's heritage in the form of existing buildings to ensure the sustainability of the resulting space.

2 Methods

A qualitative descriptive method with a historical approach. The research conducted a literature study on the concept of the Garden City in Indonesia, then conducted a field survey and also analyzed historical files such as ancient maps and wills left by colonial officials who founded Depok Lama area. The use of wills as a source of research is allowed by Creswell, this aims to be able to understand the thoughts of Cornelis Chastelein regarding the concept of what is desired for the life of the slave worker [7]. With ancient maps, identification of the development of built and unbuilt spaces is carried out, and to know the formation of space that begins with activities in the area. Then with field surveys and collection of new and old photos trying to confirm the sustainability of the space that has been formed.

3 Result and Discussion

Depok City is located in the southern part of West Java Province which is directly adjacent to DKI Jakarta Province. This economic development was followed by social developments, especially when the first State University campus moved in Indonesia, namely the University of Indonesia to Depok, then the construction of the first national housing estate in Depok City and making Depok City a buffer zone for the City of Jakarta, making Depok City a magnet for local residents [5]. The city consists of 11 sub-districts and 63 sub-districts with Sawangan being the largest sub-district with an area of 26.19 km^2. The area of Depok City is 200.29 km^2 with a population of 2,406,826 people and the density per km^2 is 12,017 people [6].

Depok Lama was formed in 1602, starting when a VOC official named Cornelis Chastelein bought private land to open a plantation business covering an area of 1244 ha. These plantations grew from mere houses for a community of free slave workers (*mardijker*) into a village.

City planning that occurred in Indonesia during the colonial period was generally carried out by the Dutch East Indies government regarding the Law of Indies document of 1573 where the completeness of the facilities and their layout was very concerned. With the issuance of the decentralization law which became the basis for the formation of the municipal system, modern concepts such as the Garden City in Indonesia emerged.

As a residential location for the slave worker community that developed into a village and residential area, Depok Lama used to have very complete facilities. There were government buildings where they managed plantation crops and houses for government leaders, schools, churches, conference halls and plantations where slaves worked. What is interesting then is that the pattern and form of this settlement is not the same as the pattern of settlements in colonial cities in general, namely the grid. Depok lama settlement pattern spread organically with residential facilities centered along the church road or what is now known as Jalan Pemuda (Fig. 1).

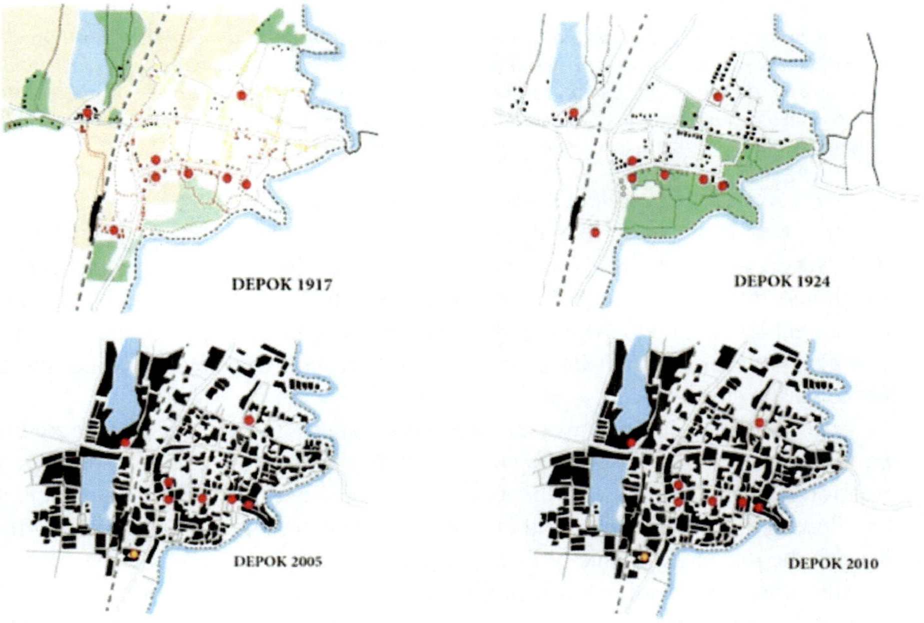

Fig. 1 Changes in the built space and the position of the building icons that tend to be the same (the author's elaboration on the old map and also the google earth map)

Chastelein died on June 28, 1714 at the age of 57 years and left a will (Het Testament van Cornelis) which contained several important messages, such as bequeathing land, money and household furniture to his daughters. Give items in the form of gold, cloth, rifles, frames, and books to Anthony Chastelein. Freeing male and female slaves, whether they are Protestant or not. Giving Depok land to Protestant slaves. However, the land may not be used for lodging or residence for Chinese and Arabs. Depok land is also not allowed to be sold and can only be used for family or religious purposes.

To make an inventory and maintain this heritage, a foundation called the Corelis Chastelein Foundation was formed in 1952. In this will, the researchers found three orders given by Chastelein to his slaves and became the background for the values of life in the social activities of citizens in shaping the space settlements.

3.1 Space to Settle

Now in Depok, 11 tenement houses have been built, namely in Ibu Negeri, and another 6 houses in Negeri Kecil near there, a total of 21 houses, all with tile roofs and the only one is enough for two houses, so 42 houses. then you can enter; there are more 3 tenement houses for Corys in Lebak near the small river and with the other 5 that's enough..

Text 1a. Chastelein's will—Order to settle [7]

... then the five or five lands that I have purchased at a price of no more than 700 Ringgit, will be owned and used by my freed slaves mentioned above along with their generations forever and ever; So they, both together, both one by one, are never allowed to sell, give to other people and pawn the land and are not allowed to transfer their rights to other people outside of them because such acts violate the law my will..

Text 1b. Chastelein's will—Order to settle [7]

From text 1a it can be seen that the two adjacent areas referred to by Chastelein Iboe Negeri and Negeri Kecil, are located on the west bank of the Ciliwung river. The first house that Chastelein built for slaves was a tile roofed house instead of tassels or coconut leaves and the walls of the house still used bamboo walls and were made of earth, which were built in mutual cooperation with Chastelein and her children borne by the costs.

According to text 1b, Chastelein wanted his slaves to stay and protect the house he built. From this mandate, the community of slaves (now called Kaoem Depok) was formed which dominates the area to this day. Based on the life values instilled by Chastelein through his will, it can be seen that Chastelein's directives indirectly formed the residential space of Depok Lama.

The house in Depok Lama with colonial architecture is a duplication of the existing building forms in Batavia, this was done as a form of social adaptation of the slaves to the government in power at that time [8].

3.2 Space to Worship

Cornelis Chastelein is a devout Protestant Christian descendant of French-Dutch, whose religious teachings are applied in everyday life in the territory that belongs to him. Apart from farming and gardening, religious activities are very important to Cornelis Chastelein.

> *… Baprima from Bali who has been baptized and given the name Lucas and who can read and write, serves and teaches these things twice in seven days to the children. I ask that once or twice every year, I will send a priest to Depok to make the holy banquet and to baptize the children and again to strengthen the belief of the people there..*

Text 3. Chastelein's will—Command to worship [7]

> *I hope that Depok will gradually become a prosperous and growing Christian community*

Text 4. Chastelein's wishes written on the memorial—Commandments to worship [7]

(Text 3) The church became the center of the activities of the slaves who occupied Depok lama residential area. Apart from studying religion, they are also taught reading, writing, and skills lessons. GPIB Immanuel Church is the first church in Depok, the initial form is simple and made of wood, then restored in 1854 to become a permanent church. Until now the church is still standing in the same location as the previous one and this is the first and oldest church in Depok. Beside this oldest church, there is a house that was used as the residence of a priest who came from Batavia which has now been converted into the office of the YLCC foundation.

The consensus on the anniversary of Chastelein's death keeps Chastelein's hope in Depok alive (text 4) and forms a continuous worship space.

3.3 Space to Study

The Depok school was formed in 1837 with students a mixture of slaves and the Bumiputera, then a Dutch school was formed in 1890 whose students were Depok people who were more 'located' and had been equated with Europeans and Europeans themselves. In his will, it is explained how Chastelein brought a teacher to teach his slaves and how to honor the teacher (text 5a and 5b).

> *Here's what I want if in the future, Mr. Pandita wants to appoint someone from outside to become a teacher in Depok, because the teacher died or for whatever reason, so that even that doesn't happen later, it will only be rejected, because black people are used to doing harm. And of course by him everyday it will become a debate, so if there is no teacher and religion is not neglected, then one person will be appointed among my people or free slaves will become a teacher, namely someone who understands the Malay language…*

Text 5a. Chastelein's will—Order to study [7]

... then even that teacher can get another quarter (1/4) and tenth (1/10) of the rice collected from the crops of the Muslims who live in Depok, and if that is not enough for the teacher, he will spend his life with his wife and their children, then the free slaves will help each year so that the teacher and his wife are not lacking and do not cause much trouble to others...

Text 5b. Chastelein's will—Order to study [7]

Currently, many public and private schools have been established in Depok Lama and schools with different religious backgrounds. The number of schools with different backgrounds becomes a diversity that complements each other and adds to the richness of the diversity that exists in Depok Lama.

Tables should be prepared using the Table function in Word preferably using the Simple 1 format. Do not include tables as graphic files or pictures.

4 Conclusions

Depok Lama in the early 1900s was a rural area that developed into a colonial village in an organic and unstructured manner and was oriented towards the welfare of its population. The formation of the Depok Lama settlement began with Chastelein's thoughts as contained in a will *(Het Testament van Cornelis)* which colored the activities of the residents, which was then reflected in the residential space and can be seen from the physical remains of the building.

Acknowledgements This research is fully funded by the Faculty of Civil Engineering and Planning, Gunadarma University, Indonesia

References

1. Huang L, Wu J, Yan L (2015) Defining and measuring urban sustainability: A review of indicators. Landsc Ecol 30(7):1175–1193. https://doi.org/10.1007/s10980-015-0208-2
2. Stelder M (2018) From the closet into the Knesset*: Zionist sexual politics and the formation of settler subjectivity. Settl. Colon. Stud. 8(4):442–463. https://doi.org/10.1080/2201473X.2017.1361885
3. Hügel S (2017) From the garden city to the smart city. Urban Plan 2(3):1–4. https://doi.org/10.17645/up.v2i3.1072
4. Creswell JW, Creswell JD (2018) Research design: Qualitative, quantitative, and mixed methods approaches, 5th edn. Los Angeles: SAGE
5. Santosa YBP, Noviyanti R (2020) Sejarah Perumnas Depok I: Perumahan Nasional Pertama di Indonesia (1974–1980). Criksetra J. Pendidik. Sej. 9(2):110–126
6. Depok BPS (2020) Kota Depok Dalam Angka 6(2)
7. Kwisthout JK (2015) Jejak-Jejak Masa Lalu Depok: Warisan Cornelis Chastelein (1657–1714) Kepada Para Budaknya yang Dibebaskan, Cetakan ke. Jakarta: PT. BPK Gunung Mulia
8. Rakhmanita E, Purwanto R, Rukayah S, Rahman A (2021) Duplikasi Bentuk Bangunan Pada Ruang Kota Kolonial sebagai Bentuk Eksistensi Diri Masyarakat Pribumi. Desain dan Konstr 1:84. [Online]. Available: https://ejournal.gunadarma.ac.id/index.php/dekons/issue/view/282

Airborne Disinfection Solution During New Normal Face-To-Face Learning

Ahmad Fitrianto, Suparlan, Askar Triwiyanto, and Lin Yola

Abstract Communities have had to make concerns when and how schools serving students in schools and campus should re-open against temporary closure to slow the spread of COVID-19 virus. Apart from continuing COVID-19 health protocols for all students, teachers and staffs, technology support needed for air circulation disinfection in crowded and inadequately ventilated space like a classroom. This study examines the impact of using UVC tools and combine with evidence-based source control practices, whether refine indoor air quality to best minimize exposure risk to COVID-19 for staff and students. This study found airborne disinfection can be solved through UVC irradiation with measured radiation 7,31 mW/cm^2 since previous published research only need 5 mW/cm^2 at least one second. The recommendation from this study, government need to raise public policy for education institutions to apply ultraviolet germicidal irradiation (UVGI) chamber to cut-off virus airborne transmission by circulating and disinfect indoor air during in-class activities all day long.

Keywords Airborne disinfection · New normal · Face to face learning · Ultraviolet germicidal irradiation · Covid 19

1 Introduction

The COVID-19 health crisis has forced hundreds of schools and universities in Indonesia to be closed temporarily to prevent the deadly virus transmission. Education institutions to make many profoundly tough choices refer to this critical situation. Communities have had to make concerns when and how schools serving students in schools and campus should re-open. This is one of the most controversial decisions

A. Fitrianto (✉) · L. Yola
School of Strategic and Global Studies, University of Indonesia, Depok, Indonesia
e-mail: ahmad.fitrianto@alumni.ui.ac.id

Suparlan · A. Triwiyanto
Faculty of Engineering, University of Indonesia, Depok, Indonesia

E. M. Nia et al. (eds.), *Sustainable Development Approaches*, Lecture Notes in Civil Engineering 243, https://doi.org/10.1007/978-3-030-99979-7_12

to be chosen. Schools have faced many difficult decisions about how to best minimize exposure risk to COVID-19 to staff and students. The results of these findings provide information on the city government's support and response to this public health crisis, and inevitably a similar crisis in the future.

2 Methods

Uncontrollable outbreaks maybe caused by airborne transmission through droplet and aerosol which permit a few viruses spread very fast among humans. Inconclusive findings frequently reported by some studies that many retrospective studies for outbreaks and transmission evidence from controlled experiments is frequently unavailable [1, 2]. However, completed investigations give possibilities that droplet and fomite transmission could also explain human-to-human transmission among case clusters [3]. So, we are confidence that SARS-CoV-2 spread by airborne transmission. This study examines the impact of using UVC tools and combine with evidence-based source control practices (i.e., masks, physical distancing, and handwashing) whether improve air ventilation and filtration through engineering controls provides a comprehensive, layered strategy to protect staffs, teachers and students from exposure to covid-19. UVC at a dose of 5 mJ/cm^2 has the potential to inactivate 99% of SARS-Cov 2 based on research statement [4]. We design and use ultraviolet germicidal irradiation (UVGI) chamber to pass air for at least 1 s with irradiation of 5 mW/cm^2. Irradiation measurement using UVC light meter. To ensure irradiation effectivity, we use bacteria specimen which have higher resistance against UVC radiation rather than SARS-CoV-2 which done by independent laboratory. Then, approaching UVGI tools as compliment facility for patients during the self-isolated period in University Indonesia alumni community.

3 Literature Review

Preventing the transmission of airborne infections requires a simple approach to inactivate the suitable aerial pathogens since the antibacterial effects of UV light in the air have long been known. UVGI can effectively inactivate both multi-resistant and drug-sensitive bacteria [5], and various viral strains [6]. In public places, traditional UVC light sources are cataracts and carcinogenic, which poses a threat to human health and therefore the spread of bactericidal UV light is very limited [7, 8]. Ultraviolet disinfection tools are mysterious to many people, how can "light" kill bacteria? The reality can verify it. The implementation of UVGI to disinfection has been a general exercise in sterile work facilities and also clinical sanitation since the mid-twentieth century. Its effectiveness has been nicely documented each scientifically and commercially. It is nature's personal disinfection/purification method.

Leading professional societies and public health institutions are publishing educational materials on addressing indoor air quality (IAQ) in schools and buildings against to COVID-19 crisis. The published guidelines converge on six indoor air quality strategies [9]. Strategies to increase fresh air through mechanical ventilation are carried out through: (a) increasing the supply of outside air through the building's heating, ventilation, and air conditioning systems; and (b) implement a flushing process between occupancy periods during which heating, ventilation and air conditioning systems run for a predetermined duration or until the purge air change target is reached. Strategies to increase outside air through the use of operable windows are pursued through: (c) opening windows to increase external flow; and (d) placement of fans in windows to circulate room air to the outside. Strategies to remove air contamination through filtration are pursued through: (e) increasing the use of air filters with a MERV of 13 or better as a target for removing airborne virus particles in the system recirculation (MERV ratings range from 1–16 with the highest value is the most efficient filter); and (f) install air purifiers with HEPA filters which efficiently capturing 99.97% of human-generated virus particles 0.3 μm in size.

School institutions make coordinated decisions on the best way to implement these six IAQ measures during a pandemic, based on current building infrastructure, financial and human resources, competing priorities, and regional policies. This study had to give it. Given the urgency of working on IAQ to ensure safe face-to-face learning, the school institutions surveyed need to understand the experience and challenges of implementing these six IAQ recommendations in the field. School institutions face challenges when trying to meet indoor air quality standards. School facilities in wealthy areas with significant leadership, as well as those upgraded before the pandemic, faced less challenges. One of the biggest challenges, called school facilities, is facilities that are not designed to support upgrades to current ventilation systems. Most school institutions face this as a major problem. Another problem is the high cost of HEPA filters. The school has been reopened and authorities are closely monitoring indoor air quality standards. Prior to the pandemic, school education institutions believed that indoor air quality monitoring was less important. Another reason they did not pay attention to air quality is due to lack of human and financial resources and the time it took to collect and monitor data. Pandemics are changing the mindset of these institutions. Today, it is important for most schools to monitor indoor air quality.

The objective of this literature review is to analyze and compare the scope and nature of the recommendations found in these documents, to note how new guidance supplements or supersedes prior design practices and to identify any significant deficiencies that should be addressed by future standards and guidelines [10]. From this point of view, the best effort suitable with Indonesia should be addressed to improve indoor air quality to support face-to-face learning in school institutions to protect staff and students from exposure to COVID-19 [11, 12].

4 Result

This study looking forward the new-normal to prepare a post-pandemic world. Investment in home infrastructure is needed to address indoor air quality (IAQ). Overall, high maintenance costs and outdated infrastructure in the face of changing climate conditions were the most cited challenges to maintain healthy indoor air quality continuously in a post-pandemic world. Recognizing the overall health and learning benefits, approaching UVGI tools as compliment facility for 100 COVID-19 patients during the self-isolated period. Those beneficiaries are University of Indonesia alumni who are infected second wave COVID-19 from June 2021 till end of August 2021. UVGI tool use UVC light as airborne disinfection method to kill or inactivate microorganisms which unable to perform replication after their nucleic acids and DNA seriously damage. Our UVC chamber design will suck air in the room through a fan and moves it towards UVC light irradiation that all microbes become killed or inactivated while passing through the chamber so that the air moving out will be microorganisms-free air.

Griffiths research statement [4] conclude that Sars-Cov-2 can be deactivated up to 90% using a dose of 5 mJ/cm^2. Thus, engineers need equipment that can pass air for at least 1 (one) second with irradiation of 5 mW/cm^2. Irradiation intensity has been measured by UV Light Meter device and the result ensures UVC tool so-called Airdisinfex effective to inactivate 99% (D99) within 0.7 s (Fig. 1). It circulates indoor air with a flow rate 150 m^3/h and irradiation within the chamber is 7.31 mW/cm^2. Microbiological tests were carried out to validate the effectiveness using test parameters in the form of bacteria. Bacteria specimen that used are Escherichia Coli, Staphylococcus Aureus and Salmonella sp. have higher resistance against UVC radiation rather than SARS-CoV-2 (Figs. 2, 3).

The results in line with testimony from all participant who used Airdisinfex during self-isolated due to COVID-19 exposure. Self-isolated person or family are generally alone with a restricted movement to each other in group or room area. Those participants feel better indoor air quality help them recovery smoothly. The existence

Fig. 1 Ceiling and standing chamber model and UV light meter result

Fig. 2 Vice Mayor of Depok City in face-to-face learning trial

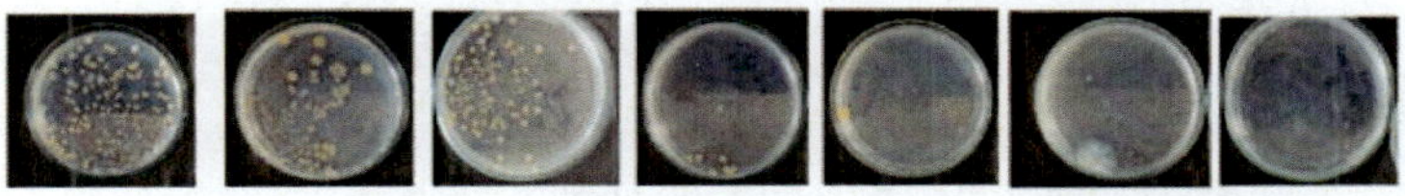

Test Sample	0 sec	1 sec	2 sec	10 sec	1 min	5 min	10 min
E. Coli (CFU/m³)	104,5	26,5	37,5	15	9,5	3	1
% Reduction		74,64%	64,11%	85,65%	90,91%	97,13%	99,04%
S. Aureus (CFU/m³)	194	175,5	141,5	105,5	4	1,5	6
% Reduction		9,54%	27,06%	45,62%	97,94%	99,23%	96,91%
Salmonella sp. (CFU/m³)	58	58	28	6,5	1,5	2	0,01
% Reduction		0,00%	51,72%	88,79%	97,41%	96,55%	99,98%

Irradiation time and % Reduction

Fig. 3 Microbacterial testing result

of Airdisinfex also give confidence to improve their immune going through a critical period of COVID-19 recovery.

5 Conclusions

Above evidence experiences address this crisis has shed light on the benefits of assuring a high level of indoor air quality for infection control and overall wellness for school institutions. Face-to-Face (F2F) learning are most similar to the traditional residential classroom experience. However, social distancing requirements will reduce the number of available seats in each classroom. Apart from continuing COVID-19 health protocols for all students and teachers, UVC light irradiation support needed to disinfect circulated air in crowded and inadequately ventilated space like a classroom. When combined with evidence-based source control practices (i.e., masks, physical distancing, and handwashing), improving air ventilation and filtration through Airdesinfex tool. It provides a comprehensive, layered strategy to protect teachers and students from exposure to COVID-19. This approach will simplify strategy related to costs and outdated building infrastructure since no need complex implementation to apply Airdisinfex both of standing or ceiling model base on every single classroom.

City Government need raise quick response to mitigate risks then make customized decision about how best to improve IAQ in all school institutions, depending on their current building infrastructure, financial and human resources, competing priorities, and local politics. Disinfection solution for airborne transmission can be addressed through UVC chamber tool which installed to sterilize circulated air in crowded and inadequately ventilated spaces. City Government need raise public policy for this public health crisis and, inevitably, future similar crisis.

Acknowledgements This paper is supported by the internal publication grant of School of Strategic and Global Studies, University of Indonesia.

References

1. Kutter JS, Spronken MI, Fraaij PL et al (2018) Transmission routes of respiratory viruses among humans. Curr Opin Virol 28:142–151
2. Tellier R, Li Y, Cowling BJ et al (2019) Recognition of aerosol transmission of infectious agents: A commentary. BMC Infect Dis 19(1):101
3. WHO (2020) Transmission of SARS-CoV-2: Implications for infection prevention precautions. http://www.who.int/news-room/commentaries/detail/transmission-of-sars-cov-2-implications-for-infection-prevention-precautions
4. Griffiths (2020) Signify and Boston University validate effecIveness of signify's UV-C light sources on inacIvaIng the virus that causes COVID-19
5. Conner-Kerr TA, Sullivan PK, Gaillard J, Franklin ME, Jones RM (1998) The effects of ultraviolet radiation on antibioticresistant bacteria in vitro. Ostomy Wound Manage 44:50–56
6. Budowsky EI, Bresler SE, Friedman EA, Zheleznova NV (1981) Principles of selective inactivation of viral genome. I. UV-induced inactivation of influenza virus. Arch Virol 68:239–247
7. Setlow RB, Grist E, Thompson K, Woodhead AD (1993) Wavelengths effective in induction of malignant melanoma. Proc Natl Acad Sci USA 90:6666–6670
8. Balasubramanian D (2000) Ultraviolet radiation and cataract. J Ocul Pharmacol Ther 16:285–297
9. Hoang A, Heming A (2021) Preparation in the pandemic: How schools implemented air quality measures to protect occupants from COVID-19. ASHRAE and The Center for Green Schools
10. Suparlan (2021) AirdisinfeX: Potentially UVC airborne disinfection from Indonesia. IUVA Asia Workshop
11. Grenhalgh T, Jimenez JL, Prather KA, Tufekci Z, Fisman D, Schooley R (2021) Ten scientific reason in support of airborne transmission of SAR-CoV-2. The Lancet. https://doi.org/10.1016/s0140-6736(21)00869-2
12. Tang JW, Marr LC, Li Y, Dancer SJ (2021) COVID-19 has redefined airborne transmission. BMJ 2021 373:n913. https://doi.org/10.1136/bmj.n913

Towards Climate Change Adaptive Spatial Planning: Urban Heat Islands Distribution in Jakarta Metropolitan Area

Sofi Ulfiasari and Lin Yola

Abstract Climate change rapidly moves due to human activities are concentrated in cities, which causes the increasing excess heat production. In addition, the ongoing development also makes the reduction in green areas that could absorb heat. Concentrated heating in the city causes an urban heat island. This study uses remote sensing analysis to measure the heat in the Jakarta Metropolitan Area using Landsat 5 and 8 imagery results in the distribution of urban heat island for ten years which has increased, both in terms of intensity and the area. To create a sustainable and liveable urban environment, it is necessary to have urban planning responsive to climate change. Government regulations to reduce urban temperatures have not been seriously discussed. On an urban scale, spatial planning policies will significantly affect the development of heat production. The finding of this study suggests scientific contributions to academia, industry, and governments informing policy framework for the UHI and climate change mitigation plan in Jakarta Metropolitan Area.

Keywords Urban Heat Island · Jakarta Metropolitan Area · Remote sensing · Sustainable city

1 Introduction

Jakarta Metropolitan Area was designed in 1950 and structured in 1952. The development center is in Taman Merdeka (Central Jakarta), surrounded by the main road and green area as a development boundary [1]. Cities can provide many socio-economic benefits. On a micro level, urban areas tend to produce more energy than rural areas. Jakarta Metropolitan Area (JMA) has six regions covering 643,789 ha, with Jakarta as the center. The Jakarta Metropolitan Area, connected by ring roads and irregularly

S. Ulfiasari · L. Yola (✉)
Department of Urban Studies, School of Strategic and Global Studies, Universitas Indonesia, Jakarta, Indonesia
e-mail: lin.yola@ui.ac.id

S. Ulfiasari
e-mail: sofi.ulfiasari91@ui.ac.id

E. M. Nia et al. (eds.), *Sustainable Development Approaches*, Lecture Notes in Civil Engineering 243, https://doi.org/10.1007/978-3-030-99979-7_13

scattered buildings, forms the phenomenon of urban sprawl. Rapid urbanization and anthropogenic warming are increasing symptoms of climate change. The development pathway with rapid urbanization has changed local to global production systems [2]. Urban production of climate risk, cities a key site within which greenhouse gases are produced [3].

Climate change affects all regions of the world, accelerating human activities in urban areas. Cities can become catalysts in global warming if not appropriately managed. Urban heat islands develop when natural land cover in an area is replaced by built surfaces that trap incoming solar radiation during the day and then re-radiate it at night [4, 5]. The configuration of open spaces and buildings in dense urban areas has a significant role in urban microclimate [6]. Walls and road materials such as asphalt are the warmer materials that can accelerate urban heating [7]. Heating in urban areas occurs shortly after sunset, while the temperature drop occurs at 3–6 p.m. The Jakarta Metropolitan Area has a diurnal air temperature variation that started to increase at 6 a.m. and peaked around 3 p.m. on all the observation points (Jakarta/Kramatjati, Bogor, Tangerang, Bekasi) [8]. The scattered observation points have different topographic characteristics but show the same results as if the heating has been evenly distributed. Economic activity is the biggest cause of accelerated temperature rise, changes in land cover make heat storage more widespread. Built-up area increasing 500% in a decade from west to east, the highest density found at North Jakarta, Tangerang and Bekasi [9].

2 Literature Review

2.1 Urban Heat Island

Urban Heat Island (UHI) is defined as the temperature increases in urban areas compared to the surrounding environment after sunset [10–13]. Meteorological, Climatological, and Geophysical Agency (locally known as BMKG) defines an urban heat island as a metropolitan area characterized by a higher ambient temperature compared to the surrounding non-urban regions or suburban areas, the cause of which is the higher absorption of solar energy with building materials in urban areas made of materials such as glass, asbestos, concrete, asphalt [14]. UHI is related to urban energy balance in the form of thermal energy flow; it is composed of solar radiation, heat storage, heat generated by human activities, sensible heat, and latent heat, which can also be converted into other forms of energy to move in the surface flow [15].

2.2 *Urban Sustainability*

Urban sprawl is not the result of explicit government policies but the somewhat inexorable product of mobilization [16]. Urban sprawl is considered an important problem for sustainable living. The sustainable city concept produces urbanization and urban growth direct to environmental problems. Sustainability has spawned much analysis and many different definitions [17]. Contain in SDG's Goal11, because the future will be urban for most people, the solutions to some of the most significant issues facing humans-poverty, climate change, healthcare, and education must be found in city life [18].

Multilevel governance plays a vital role in taking a step toward climate change and urban sustainability [3]. According to the Fifth Assessment Report of the Intergovernmental Panel on Climate Change (IPCC), three categories of adaptation options can be identified, namely structural/physical, social, and institutional [19]. Control of land use change refers to spatial planning, the need for land protection from a climate adaptation perspective, which is enabled by all types of plans for the wider area [20]. However, the urban spatial studies has been associated to range of variables to investigate the best strategies and solutions for planning a more sustainable cities [21, 22].

3 Methods

This study was conducted in Jakarta metropolitan area (locally known as Jabodetabek). Using remote sensing technology as a research method, a whole scene Landsat 5 and 8 (path 122 rows 64 and 65) of 2009/07/29, 2014/9/13, and 2019/8/1 was acquired with less cloud cover. The formula for calculating the urban heat island intensity was reviewed from the following reference [23]: The threshold value of urban heat intensity expressed in μ as average surface temperature and α as standard deviation and heat island intensity formula [24, 25]. T_{rad} represents the surface temperature.

$$\Delta T = \left(\mu + \frac{\alpha}{2}\right) \underline{\text{and}}\, I_{HI} = T_{rad} - \Delta T$$

4 Results

Rapid development has occurred in the last few decades. In 2009, the distribution of urban heat islands in Jakarta was centered in DKI Jakarta to surrounding areas like District of Bogor, Kota Bogor, Tangerang Selatan, and Tangerang. The UHI intensity that dominates Jabodetabek ranges from 1 to 4 °C; areas with high intensity are seen in the center of Jakarta. Parts of the Bekasi Regency show the urban heat island even though the site is a suburb. This condition indicates that the development of the city towards the sub-urban is enormous so that it affects a significant modification of the atmosphere. The growth in 2009 had a substantial effect on the urban climate; warming occurred not only in DKI Jakarta but almost all areas of Bekasi and Tangerang also have the same characteristics.

In 2014, the distribution of heat islands in Jabodetabek showed that the condition of heat islands at 5 °C occurred a lot. A road network connected DKI Jakarta to Depok City, Bekasi City, was spread unevenly, and parts of Bekasi Regency showed heat island intensity up to 8 °C. The increase in area in the Bogor City area is located at a higher altitude than Jakarta but has the exact characteristics of the urban heat island. UHI spreads in the outermost region of Greater Jakarta with a value of 1 °C. Indications of UHI are also closely related to the physical development in that location.

The distribution of UHI in 2019 is shown in Fig. 1, which emphasizes the increasing damage to the atmosphere and the environment. The UHI value in the center starts at 5 °C, which is very high in contrast to the surrounding area. These areas form a pattern connected, covering the face of the city of Jakarta to the city of Bekasi. A small area with a UHI value of 1 °C in 2014 merged to form a large area with a heat island value reaching 4 °C in 2019. Spatially, the increase in size, more even distribution, and the growth of the new urban heat island make Jabodetabek a dangerous metropolitan area. The heating that occurs continuously until it penetrates mountainous areas makes it necessary to evaluate development to reduce the rate of urban warming.

Spatially, the increase in the distribution and value of UHI that has occurred in Greater Jakarta for a decade is evenly distributed, especially in areas directly adjacent to DKI Jakarta. During the first five years, the size of the UHI of heat decreased by 11,399 ha and increased considerably in the next five years by 54,796 ha. Over a

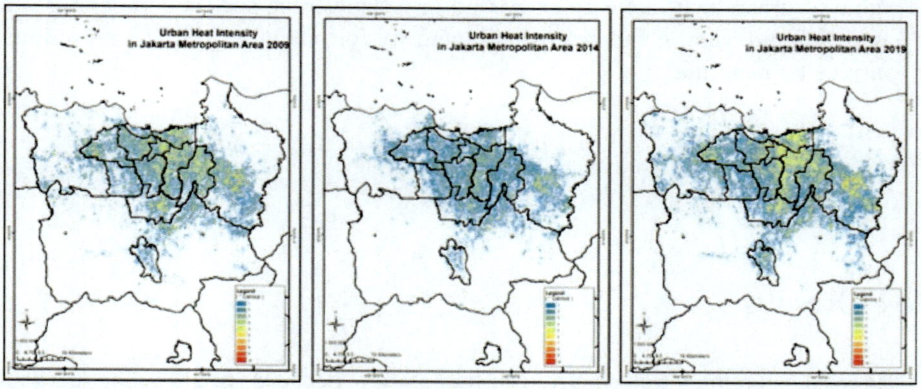

Fig. 1 Urban heat intensity in Jabodetabek on year of 2009, 2014 and 2019

Table 1 Urban heat Island intensity and comparison of area among 2009, 2014 and 2019

Urban Heat Island Intensity (°C)	2009 (ha)	2014 (ha)	2019 (ha)
1	51,671	69,517	54,860
2	40,494	48,546	64,007
3	36,404	157,2	39,380
4	14,963	2451	22,498
5	4220	520	1657
6	492	148	8874
7	114	48	351
8	58	12	132
9	12	5	50
10	2	1	17
Total	148,430	137,031	191,827
UHI of Jakarta Metropolitan Area	23,05%	21,28%	29.80%

decade, the increase in the area and the rise in temperature values. This can lead to changes in local climatic conditions.

The urban heat island reached 14,843 ha in 2009 with the largest area of 1 °C. The higher temperature difference with the lower area. The urban heat island was found in an industrial location, precisely in a factory in Bekasi Regency, dominated by concrete and buildings with heat-absorbing materials and the absence of green areas. In 2014, UHI intensity range four and 5 °C, while at a discount of 1 °C, the area increased to 18,146 ha. This increase in the area indicates a new site experiencing a growth in excess heat. During the 2009–2019 period, the largest area of 1 °C heat island value was seen in 2014, 69,517 ha or about 10.9% of the Jabodetabek area. In 2019 being the warmer year for ten years, the urban heat island experienced an increase in temperature and area. Even the difference in temperature between urban and sub-urban areas of 6 °C has an area of 1.3% of the Jabodetabek area. The difference in temperature above 6 °C spreads in Bekasi Regency and Depok City. The development of industrial and residential areas is the main trigger for increasing heat islands.

Presidential Decree No. 54 of 2008, Article 32 paragraph 4 states that the proportion of urban/urban public green open spaces is at least 20% of the area of each Jabodetabek city. Completing Presidential Decree No. 54 of 2008 to become Presidential Decree No. 60 of 2020 is the first step to emphasize regional integration. In line with Presidential Decree No. 60 of 2020, Bodetabek is designed as a residence/settlement. There is a buffer and protective area expected to support the burden of Jakarta's conurbation, as is the case in providing green space to reduce the impact of warming in the city. In 2014 the intensity of the heat island above 5 °C began to spread evenly and from the west–east direction through Jakarta. Steps that can be taken to reduce the island of heat are adapting Presidential Regulation No. 61 of 2011 concerning the national action plan for reducing greenhouse gas emissions, setting out mitigation steps that can be taken by each region, including the implementation of environmentally friendly transportation development, especially in the DKI Jakarta, Banten Province, and Banten Provinces. West Java so that all three are connected without producing high exhaust emissions.

The action to reduce the intensity of the heat island in Indonesia has not been regulated in legislation; no regulation governs the intensity limit of the island of heat that is not harmful to the community. The integration of the Jabodetabek area arrangement is not in line with the plans that have been made; the phenomenon of the heat island, which cannot be denied, is in the zone of the designated area to be built so that efforts to reduce the heat island by providing green open space in the site are not accessible. In addition, the Presidential Regulation does not contain a single paragraph that discusses the impact of environmental damage caused by air damage. The results of this study are expected to be helpful and contribute to the formation of new regulations.

5 Conclusions

The distribution of the heat island in Jabodetabek is strongly related to the road network infrastructure development. The heat stress developed along the densely built areas in 2009 was seen in the DKI Jakarta area and some others buffer zones such as parts of Depok City and Bekasi Regency with an area of 148,430 ha of the heat island. In 2014 the pattern of heat islands developed from the city center to buffer zone with an area of 137,031 ha. Meanwhile, in 2019, the heat island covering an area of 191,827 ha was distributed evenly from the upstream, middle, and downstream regions; conditions seemed un-control considering that the heat island that occurred covered up to 30% of the entire Jabodetabek area.The implication of the island of heat on the direction of development is the cycle of increasing urban heat that continues to increase due to the rapid city development. In response to this, the authority suppresses the heat island mitigation through the regulation of PP No. 60 of 2020 and PP No. 61 of 2011 to be implemented by all administrative area.

References

1. Rukmana D (2008) Planning the megacity: Jakarta in the twentieth century. J Am Plann Assoc 74(2):263–264. https://doi.org/10.1080/01944360801940995
2. Lundqvist M (2007) Sustainable cities in theory and practice: A comparative study of Curitiba and Portland. Karlstads universitet
3. Bulkeley H (2013) Cities and climate change. Routledge, London
4. Quattrochi D, Luvall J, Rickman D, Estes M, Laymon C, Howell B (2000) A decision-support information system for urban landscape management using thermal infrared data. Photogramm Eng Remote Sens 66(10):1195–1207
5. Oke TR (1982) The energetic basis of the urban heat island. J R Meteorol Soc 108(455):1–24

6. Yola L (2018) Impact of urban configurations on microclimate and thermal comfort in residential area of Kuala Lumpur
7. Arifwidodo S, Chandrasiri O (2015) Urban heat Island and household energy consumption in Bangkok, Thailand. Energy Procedia 79:189–194. https://doi.org/10.1016/j.egypro.2015.11.461
8. Widysamratri H, Souma K, Suetsugi T (2019) Study of urban temperature profiles on the various land cover in the Jakarta metropolitan area. Indonesia. Indonesian Journal of Geography 51(3):357–363
9. Ulfiasari S, Yola L (2022) How does urban development contributes to urban heat Island: A decade increase of urban heat intensity in Jakarta metropolitan area. In: Yola L, Nangkula U, Ayegbusi OG, Awang M. (eds) Sustainable architecture and building environment, vol 161. Lecture Notes in Civil Engineering, Springer, Singapore. https://doi.org/10.1007/978-981-16-2329-5_9
10. Erell E (2012) Urban microclimate. Routledge, In Urban microclimate. https://doi.org/10.4324/9781849775397
11. Oke TR (1988) The urban energy balance. Prog Phys Geogr 12(4):471–508. https://doi.org/10.1177/030913338801200401
12. Van Bohemen H (2012) (Eco)system thinking: Ecological principles for buildings, roads and industrial and urban areas. In: V. H. van Bueren E, van Bohemen H, Itard L (eds) Sustainabl). Springer, Dordrecht. http://doi-org-443.webvpn.fjmu.edu.cn/10.1007/978-94-007-1294-2_2
13. Voogt JA, Oke TR (2003) Thermal remote sensing of urban climates. Remote Sens Environ. https://doi.org/10.1016/S0034-4257(03)00079-8
14. Badan M, Klimatologi dan G (2013) Kamus Istilah Perubahan Iklim. In Researchgate.Net. available in: https://www.researchgate.net/profile/Edvin_Aldrian/publication/305358957_kamus_istilah_perubahan_iklim/links/578a14df08ae59aa667932ac/kamus-istilah-perubahan-iklim.pdf
15. Harman IN (2003) The energy balance of urban areas. The University of Reading Department of Meteorology T, pp 1–169
16. Glaeser EL, Kahn ME (2005) Sprawl and Urban growth. SSRN Electron J. https://doi.org/10.2139/ssrn.405962
17. Goode N (2011) Sustainable cities (A Vision of Our Future Landscape). Grant Thornton
18. The Sustainable Development Goals (2018). United Nations. Available online: https://www.un.org/sustainabledevelopment/wp-content/uploads/2018/09/Goal-11.pdf
19. Intergovernmental Panel on Climate Change (IPCC) (2019) Special report on climate change, desertification, land degradation, sustainable land management, food security, and greenhouse gas fluxes in terrestrial ecosystems summary for policymakers. Available online: https://www.ipcc.ch/srccl/ApprovedDraft, https://www.ipcc.ch/srccl
20. Thoidou E (2021) Spatial planning and climate adaptation: Challenges of land protection in a Peri-Urban area of the mediterranean city of Thessaloniki. Sustain 13:4456. https://doi.org/10.3390/su13084456
21. Yola L, Ho CS (2017) Computer simulation as an alternative approach in climatically responsive urban configuration study. Chem Eng Trans 56(2017):505–510
22. Yola L, Ayegbusi OG, Saiya HG, Djaja K (2022) Empirical and numerical approaches in urban microclimate modeling: Investigation on the reliability. Lect Notes in Civil Eng (161):51–57. https://doi.org/10.1007/978-981-16-2329-5_7

23. Ma Y, Kuang Y, Huang N (2010) Coupling urbanization analyses for studying urban thermal environment and its interplay with biophysical parameters based on TM/ETM+ imagery. Int J Appl Earth Obs Geoinf 12(2):110–118. https://doi.org/10.1016/j.jag.2009.12.002
24. Alves EDL, Lopes A (2017) The Urban heat Island effect and the role of vegetation to address the negative impacts of local climate changes in a small Brazilian city. Atmosphere 8(2):18. https://doi.org/10.3390/atmos8020018
25. Ozdemir H, Unal A, Kindap T, Turuncoglu UU, Durmusoglu ZO, Khan M, Tayanc M, Karaca M (2012) Quantification of the urban Heat Island under a changing climate over Anatolian Peninsula. Theoret Appl Climatol 108(1–2):31–38. https://doi.org/10.1007/s00704-011-0515-8

User Statements of Cibitung—Cilincing Toll Road Route Operation During Pandemic Covid-19

Pamuka Prasetya Primantara and Lita Sari Barus

Abstract The purposes of this research are to determine the potential and usefulness of Cibitung Cilincing toll road roles during pandemic covid-19. Despite regulation of movement restrictions in Indonesia, the Bekasi district area residents also demand efficient time & cost terms of their mobilities. This research uses the theory of city access concepts and urban mobility. The author used the survey method, a literature study. This research is expected to explore and measure the roles of Cibitung Cilincing Toll Road to shorten mobilities distance and more affordable costs incurred with general comparison before and after the Cibitung Cilincing Toll Road partially operated 2,75 km in August 2021. The results of this study will describe the user statements considered and related to cost, time, and their characteristics. Increased mobility in this area will undoubtedly gain potential regional development from a social perspective.

Keywords Cibitung Cilincing Toll Road · Mobilities · User statements · Digital questionnaire · Covid-19

1 Introduction

Covid-19 pandemic hit Indonesia in all sectors, we are still struggling to improve the situation and conditions in all aspects, not only in terms of vaccines but also to look after the people infected the Covid-19. Economic and business activity also rapidly decreased from march during pandemic. Therefore, the government also needs to improve the community's mobility so that it returns to normal. In the past, the public has learned that there are some regulations and restrictions on people's movement and mobility to prevent the spread of Covid-19, from March 2020 until September 2021. The government has also started exploring new standard policies. The movement of people is associated with the efficiency of travel costs, and congestion is often

P. P. Primantara (✉) · L. S. Barus
Department of Urban Studies, School of Strategic and Global Studies, Universitas Indonesia, Kota Depok, Indonesia
e-mail: pamuka.prasetya01@ui.ac.id

L. S. Barus
e-mail: litabarus@ui.ac.id

E. M. Nia et al. (eds.), *Sustainable Development Approaches*, Lecture Notes in Civil Engineering 243, https://doi.org/10.1007/978-3-030-99979-7_14

one of the reasons obstacles. People want fast and smooth access to reduce their mobility time during this covid-19 pandemic and minimize costs. The construction of the Cibitung toll road to Tanjung Priok Port is part of the government's mandate to realize the 2015–2019 Medium-Term Development Program, which provides road infrastructure to support the national economic growth strategy.

2 Literature Review

The concept of the city associated with access Kevin Lynch can be classified according to the features/what is provided by the access and to whom the access is granted. The most basic is probably access to other places in the sense of the word community movement: Access to relatives, to friends, to business partners, and various other networks. Humans are social creatures, and in essence, will often communicate and will regularly meet their relatives [2]. At least one day a week, there is a meeting between members of a social group, and according to Kevin Lynch, this need is a basic need for the welfare of society. Primitive societies also classify their places of residence according to this rule, and so do modern societies. However, electronic communication can develop and nowadays almost replace physical proximity to digital communication. Nevertheless, travelling to visit other people is still an essential component of travel and access for urban communities. What is also important is the community's access to their activities. The primary activities for many adults may include work and housing, but we must also have essential services such as activities; financial, health, recreational, educational, and religious activities. This activity is essentially an opportunity for the person to survive, such as work, worship, study, or recreating the harmony of community life, for example. During the pandemic Covid-19 condition, this theory is also essential for accessing health facilities, particularly in emergency conditions such as roads to hospitals. However, the most significant number of recorded city trips are still primary: business trips, going from home to work, transportation of goods and services, and others. As stated by Frederic Stout & Le Gates, in City Reader; Sixth Edition, which is about new urban mobility [3]. First, the impact of mobile technology on critical aspects of city life, such as the existence of legal and government/regulatory functions in this case, such as the existence of a market function, the part of economic production and trade, as well as the role of the individual community, family, environment, and local culture. Second, the influence of globalization in urban areas on aspirations that arise due to the mobility of the two main constituencies that will become new policies in future technology in developing countries (middle class) and countries with "millennial" residents, namely those born in the 90s. And the third, is the ways that will emerge in the digital era, and communication technology and urbanization itself which can process over time. And will most likely answer the challenges of population growth, demographic bonuses, and social equality that will shift to the urban urbanization paradigm.

The new geographic, economic theory emphasizes circular causality mechanisms to explain the spatial concentration of economic activity [4]. In this model, the centripetal strength comes from the variation in consumption or the variety of intermediate goods on the production side. The centrifugal force stems from the pressure exerted by the geographic concentration of local input markets that offer higher prices and spread demand. If transportation costs are low enough, aggregation will occur. In the technology externality model, knowledge transfer between firms provides an incentive for the aggregation of economic activity. Because these interactions are informal, the extent of information exchange decreases with increasing distance. This incentivizes entrepreneurs to be located close to other companies resulting in agglomeration [5].

3 Research Method

The research method that the author uses in this research is the survey method. The survey research method is research in which the primary source of data and information is obtained from respondents as research samples using or questionnaires as data collection instruments. Individuals are the sample utilized as the unit of analysis in most cases. Other branches of study, such as homes, groups, enterprises, and the state, can be used. The author attempts to use this survey approach as one that is based on accurate data and avoids terrible data, such as data from unreliable websites. When conducting survey research, one thing to remember is that samples are used as the primary data source. According to Darmadi, conducting this literature study is carried out by researchers after they determine the research topic and the formulation of the issues before they go into the field to collect the data needed in research as evidence in conducting analysis and making conclusions on a desired study or research. academic-based [6]. Assignment (Figs. 1, 2, 3, 4, 5, 6, 7, 8, and 9).

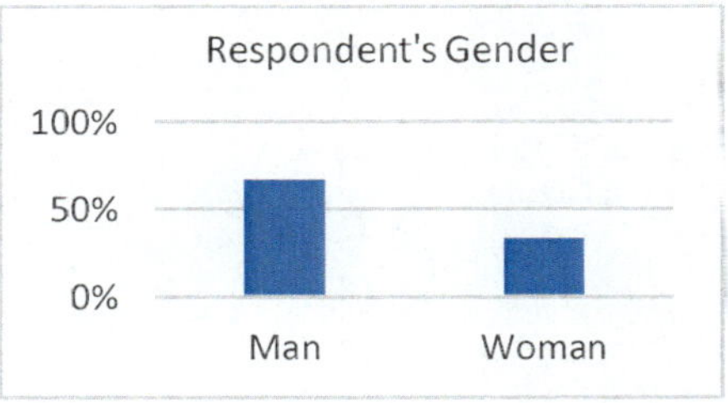

Fig. 1 Respondent's gender

Fig. 2 Respondent's age

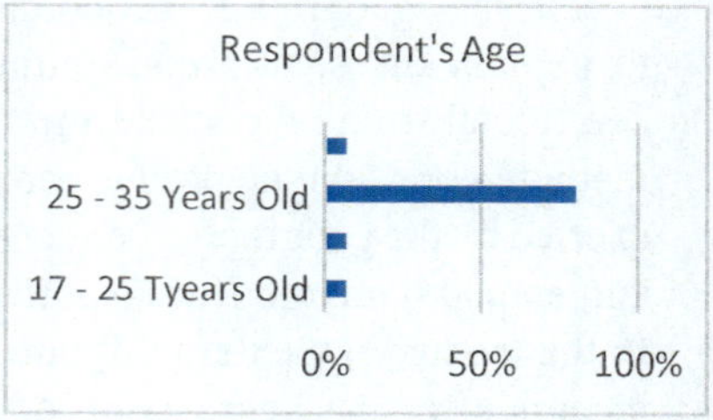

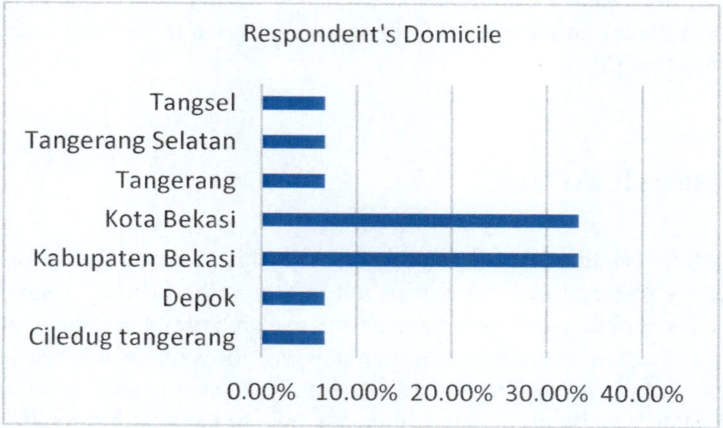

Fig. 3 Respondent's Domicile

Fig. 4 Respondent's satisfaction

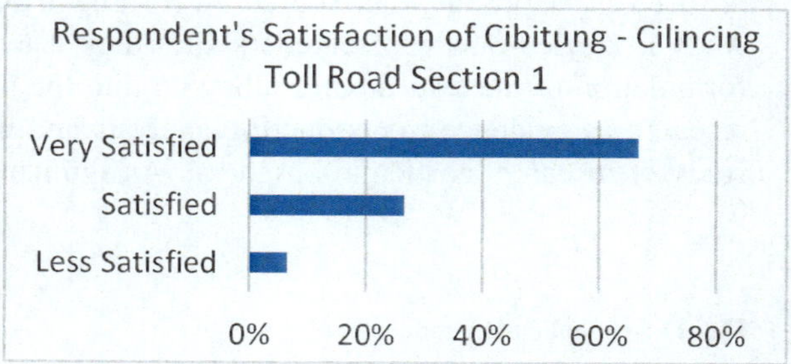

Fig. 5 Respondent's cost satisfaction

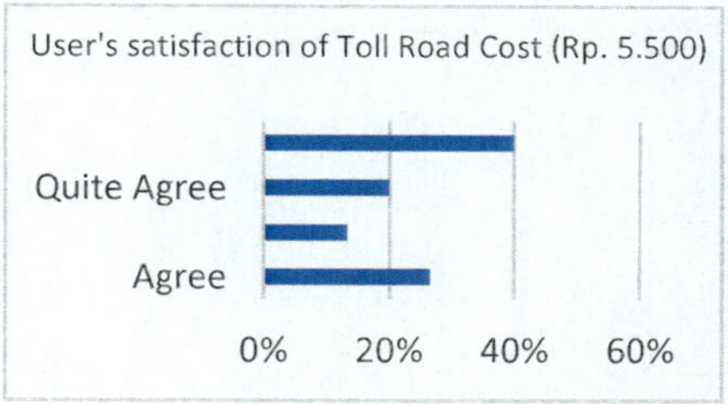

Fig. 6 Respondent's work location

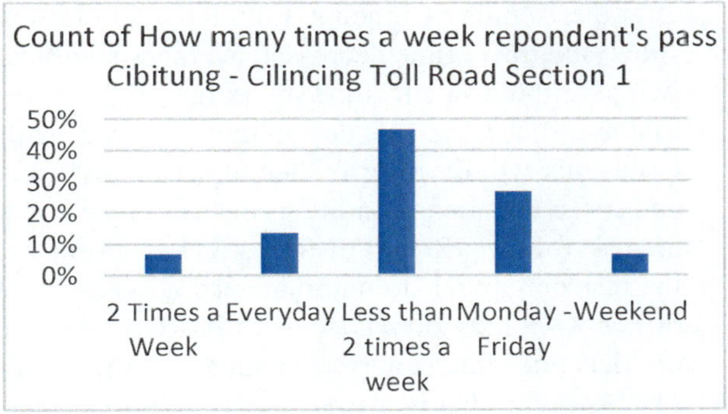

Fig. 7 Respondent's times passes Cibitung–Cilincing Toll Road

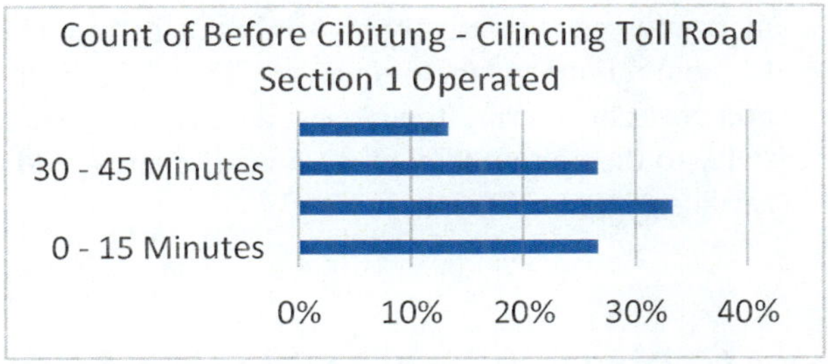

Fig. 8 Respondent's travel time before Cibitung–Cilincing Toll Road operated

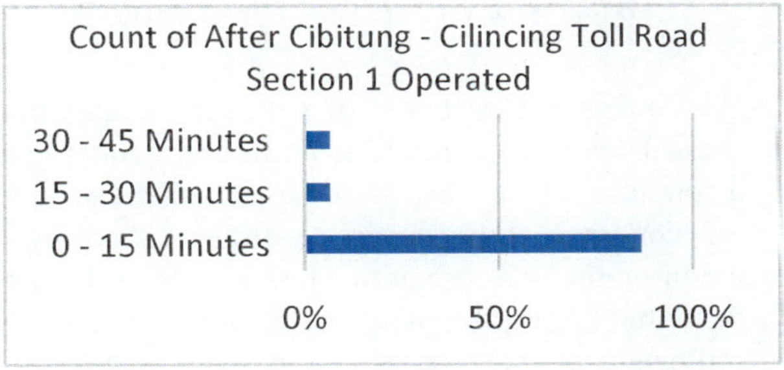

Fig. 9 Respondent's travel time after Cibitung–Cilincing Toll Road operated

4 Results and Discussion

Regarding the survey results, the author doing the bench-marking analyses and expected to get a comparison between; Before Cibitung Cilincing Toll Road Operation from Cibitung Junction to Telaga Asih Junction During covid-19 pandemic. After Cibitung Cilincing Toll Road Partially Operation from Cibitung Junction to Telaga Asih Junction during the covid-19 pandemic.The result of this survey was average travel time before the Cibitung Cilincing toll road operation was 30–45 min, and after the Cibitung Cilincing, toll road operations decreased to 0–15 min. The result shows the travel time decreased averages significantly from 15–30 min decrease. It will save the residents time mobilities, and it is the more affordable cost compared to non-toll access roads during the movement with their vehicles. There are several issues that will discover related to access and connectivity that will arise with time and cost comparison conditions in terms of access inter-regional in the Bekasi district area, as follows: Bekasi district area has several industrial & commercial areas, and the residential area around that need good access & mobilities. Bekasi district area does not yet have direct access to JORR 1 and access to Tanjung Priok Port. Bekasi district area, which intersects with the Cibitung Cilincing Toll Road, only has a local frontage road that is also across the national road, which the capacity has limited road width. In terms of mobilities frequency of the areas, Bekasi district residents mobilities were relatively high in intensity, mainly related to the economic level of the community and the frequency. Therefore, the mobility requires considerable time and cost considering the closest access to the surrounding Toll Roads and connectivity to the commercial area, education area, and public facilities, Hospital, public services, station, shelter.

4.1 Potential Connectivity Around Bekasi District Area: Goods & Services Flow Benefits

The unequal distribution of resources creates disparities in the rate of economic growth between regions. This resource inequality is reflected in the concentration of economic activity that occurs in certain areas only. Areas, where the concentration of economic activity occurs benefit are called agglomeration economies. Agglomeration economy is an external resulting from the geographical proximity of economic activities [7]. Furthermore, the existence of an agglomeration economy can positively influence the rate of economic growth. As a result, areas included in agglomeration generally have a higher growth rate than those not agglomerated. The positive relationship between geographic agglomeration of economic activities and economic growth has been widely demonstrated is in line [8]. Agglomeration produces spatial differences in income levels. The more spatially agglomerated an economy will increase its growth. Areas with many processing industries grow faster than areas with few processing industries. The reason is that regions that have more processing industries have accumulated capital. In other words, areas with a concentration of processing industry grow faster than areas without the attention of the processing industry. Given the facts above, this study will analyze the impact of aggregation on the economic growth of Bekasi Regency as the impact caused by the construction of the Cibitung Cilincing Toll Road.

4.2 Inter-regional Benefits in Bekasi District Area

4.2.1 Inter-regional Connectivity Potential of Bekasi District Area

The Cibitung Cilincing Toll Road was built in the National Strategic Project (*PSN*: *Proyek Strategis Nasional*) [9]. In Bekasi Regency itself, the Cibitung Cilincing Toll Road passes through the following areas at the toll gate area. Transportation of goods, logistics, and private vehicles will go to the Bekasi district, particularly Cibitung, Tambelang, & Tarumajaya areas. Of course, it will be easier if the Cibitung toll road is completed and immediately operated from all sections. Among the regions mentioned above, particularly in Bekasi Regency, many MSMEs, small industries, logistics centers, housing, educational facilities, public facilities, and local government offices will provide mutual benefits, particularly road connectivity. A toll road will significantly impact travel time and regional development from economic, social, and environmental aspects in terms of connectivity. The area that is passed is an area that, from the author's view, it will have a significant impact due to the passage of the Cibitung Cilincing Toll Road, particularly in the toll gate area, which allows vehicles to enter and leave, and it is hoped that there will be economic development and impact on the surrounding community, including; Support the emergence of new MSME around toll gates, Support the emergence of new residential land due to access to toll roads, Support the emergence of new commercial areas around toll gates, The emergence of new health facilities & public health facilities due to many new residential areas, The increased flow of goods & services passing through the area. Congestion is one of the urban issues, particularly in big cities in Indonesia in general and particularly in agglomeration areas such as *Jabodetabek & Bekasi* Regency areas, in this case, the places that I researched. The flow of traffic passing by daily vehicles determines how smooth traffic is, the volume that occurs, and the condition of the existing roads. When it comes to non-toll roads, many factors affect congestion in the area, including Traffic Lights [10].

5 Conclusion

The distribution of goods and services is one of the main things in regional development and the economic development of a region, particularly during pandemic covid-19. Therefore, the development of the area around the toll road can also grow the regional economy so that the flow of goods and services that are not disrupted due to traffic factors can minimize additional costs due to the disruption of the flow. In addition, commodity prices remain stable and follow market conditions not to increase fees in obtaining goods and services. Traffic factors and also delivery time influence the flow of goods & services. Therefore, time and smooth traffic are the main supporting factors to distribute commodities properly by using toll roads/expressways. Moreover, JORR II has been connected by the toll road to Tanjung Priok Port. Therefore, with the opening of the Cibitung Cilincing Toll Road, the flow of loading and unloading goods, exports and imports, and others will be faster with each other from and to the Tanjung Priok Port. The existence of toll road construction has affected the distribution of development results and justice and the surrounding economic development. The toll road itself belongs to the upper class and big industrial side and the benefit of people from all walks of life, including those felt by the people of Bekasi Regency, the Cibitung Cilincing Toll Road passes. Moreover, Cibitung Toll Road can move the economy around its area. As follows: Opening Jobs for the community around the construction of the Cibitung Cilincing Toll Road, provide access to residential areas to reduce transportation costs, providing light for new MSMEs that will open their businesses around Toll Roads, Integrated rest areas/Rest Areas can become new land & business opportunities for the surrounding community, Reducing the level of congestion so that travel time in work and activities becomes shorter.

References

1. PT Cibitung Tanjung Priok Port Tollways (n.d.) CTP tollways. http://Ctptollways.Co.Id/. Retrieved July 2, 2021, from http://ctptollways.co.id/
2. Lynch K (1981) A theory of good city form. MIT Press, Cambridge, Mass
3. LeGates RT (2011) The city reader. In The city reader. https://doi.org/10.4324/9780203869260
4. Krugman (1998) Space: The final frontier. J Econ Perspect 12(2):161–174
5. Jamzani S, Dedi I, Aglomerasi dan Pertumbuhan Ekonomi: Peran Karakteristik Regional di Indonesia
6. Kartiningrum E (2015) Panduan Penyusunan Studi Literatur
7. Bradley R, Gans JS (1996) Growth in Australian cities. The Economic Record. Econ Soc Aust 74(226)
8. Martin P, Ottavianno (2001) Growth and agglomeration. Int Econ Rev 42(4):947–968
9. Peraturan Menteri Koordinator Bidang Perekonomian Republik Indonesia Nomor 7 Tahun 2021 Tentang Perubahan Daftar Proyek Strategis Nasional
10. Kementerian Pekerjaan Umum dan Perumahan Rakyat (n.d.) BPJT—Badan Pengatur Jalan Tol. https://Bpjt.Pu.Go.Id. Retrieved July 2, 2021, from https://bpjt.pu.go.id/konten/jalan-tol/sejarah

Marine Pollution Challenges in Jakarta Northern Coast Fishermen Settlement

Lin Yola, Sofi Ulfiasari, Garrin Alif Nanditho, Rosnaini Daga, Mulawarman Hannase, and Nur Fatwa

Abstract Indonesian coastal cities grow significantly with the strategic economic value. Jakarta is no exception. As the capital city and metropolitan, with the complexity of social and environmental long-term issues, the rapid growth of city economy and development of Jakarta coastal area and generates the big impact on social and environment. The marine pollution in Jakarta Northern Coast area causes the environmental and social issues in the informal fishermen settlement. The local communities were forced to adapt with these challenges, as there was limited affordable living environment alternative available for the poor. Government has planned the long-term agenda to mitigate the risks and the better future for the local community and environment. There was gap between the ground challenges faced by the fishermen community and the government related policies in achieving the better coastal development. This study presents a case-based analysis in Jakarta Northern Coast Fishermen Settlement. Results show that the multilayer backgrounds of social, economy and governance roles are complex interrelated factors to overcome the long term social and environmental challenges. A holistic approach is suggested for a strategic solution for a more sustainable and resilient Jakarta Northern Coast Area.

Keywords Jakarta Northern Coastal · Fishermen settlement · Social and environmental challenges · Marine pollution

L. Yola (✉) · S. Ulfiasari · G. A. Nanditho
Department of Urban Studies, School of Strategic and Global Studies, Universitas Indonesia, Jakarta, Indonesia
e-mail: lin.yola@ui.ac.id

R. Daga
Faculty of Economics, STIM Nitro, Makassar, Indonesia

M. Hannase
Department of Middle East and Islamic Studies, School of Strategic and Global Studies, Universitas Indonesia, Jakarta, Indonesia

N. Fatwa
Department of European Studies, School of Strategic and Global Studies, Universitas Indonesia, Jakarta, Indonesia

E. M. Nia et al. (eds.), *Sustainable Development Approaches*, Lecture Notes in Civil Engineering 243, https://doi.org/10.1007/978-3-030-99979-7_15

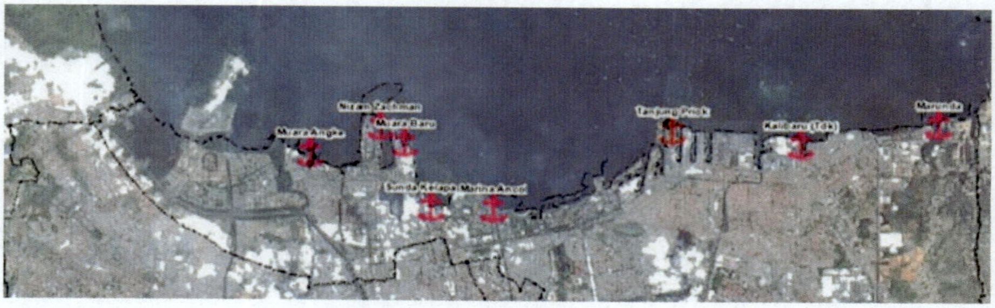

Fig. 1 Jakarta coastal area

1 Jakarta Coastal Area

Pasundan Kingdom, the origin of Jakarta, well-known as its' great era in trading port in coastal area has been explored till today rapid development time [1]. Jakarta's coastal area (Fig. 1) with a 514 km^2 resources play an important role in its' the economic movement. The coastal area has been also developed for a multi-use such as tourism, fishery, trading and a high traffic international port. The port areas are inseparable part of the fishermen settlement, as almost all fishermen settlement ranging from the informal to the formal ones are situated around the ports to support the economic activities. The rapid population and development have caused the increase of built-up land in the coastal area. This condition leads to the disposal of waste generated by the surrounding residents' activities [2] that cause damage of marine environment and coastal settlement, including the marine pollution.

The government through a spatial plan has regulated the management and land use along the coastal area of Jakarta Northern Cost. However, the planned plan faced some challenges to be enforced, thus it needs the support of the community for a holistic and resilience progress. This study explores two fishermen informal settlements within Jakarta Northern Coast area to investigate the social and environmental challenge in marine pollution.

2 Jakarta Marine Pollution Situation

Marine pollution has been discussed widely in Indonesian's environment debate. In general, the marine pollution has been resulted by the poor in land waste management. The Coordinating Ministry for Maritime and Investment Affairs [3] reported that 1.29 million ton yearly plastic waste pollute the coastal areas. A study [4] reported the accumulating Indonesian's coastal plastic debris has reached up to 0.40 Mt/year. Strategic policies and actions were intensely developed in addressing this challenge. At national level, Indonesia has released Presidential Decree Number 83/2018 on commitment to mitigate the marine debris. It aims to target the marine debris mitigation to 75% in 2025. It also presents the National Action Plan (2018–2025) with the strategic marine mitigation agenda.

Coastal border areas have the potential to withstand the potential of land subsidence. In its quest to solutions to abrasion, land subsidence, flood, environmental degradation due to rapid development, the local and national authorities have created the National Capital Integrated Coastal Development (NCICD) masterplan in form of giant sea wall [5]. The seawall aimed to protect more than half of Jakarta's coastal communities from the high tide issues following the climate change symptom. However, the fishermen complained that the seawall cause the reduce of fishes around the coastal as the seawall that stretches making the marine ecosystem distance from the mainland. The sea wall and the pollution along the Jakarta Northern Coast area are the interrelated variables directly impact the social and environmental issues for the coastal communities.

3 Method

This study presents the impact of marine pollution on social environmental sectors in two informal fishermen settlements along Jakarta Northern Coast areas; Kalibaru and Muara Baru. Climatically, the case area is a hot and humid climate with the air temperatures range up to 22–37 °C. The land subsidence recorded between 1 to 15 cm every year. This area was recorded with the highest poverty compared to other areas in Jakarta [6]. Some of fishermen live on the water with their boats. The various economic and development activities in Kalibaru and Muara Baru such as real estate, power plant, industrial area, warehousing area, and reservoir cause a high volume of waste disposal and development of marine pollution. It was recorded that 63.6% of the area heavily polluted in 2015.

The primary data collection of this study was conducted through the field observation and deep interview on the fishermen in the two case studies. The literature review on the relevant policies and studies on the marine pollution issues and strategies were also conducted. The investigation on the marine pollution particularly the solid waste challenges were focusing mostly on the causal analysis of the pollution and social environmental aspects in the fishermen communities. The qualitative descriptive analysis of the primary and secondary data was used to present the research argument.

4 Marine Pollution and Coastal Communities

Coastal areas were mainly targeted for continues development due to its' richness of natural resources, energy sources, and ecosystem services, as well as range of socioeconomic potential features [7]. The marine pollution as the results of the massive coastal development causes the multiple impacts in coastal social, economic and living environment. The concentration of plastic in the water threatens marine life and fishing activities [8]. Thus, the marine pollution control should not only consider the marine ecological life but also the economic consequences [9]. The economic loss due to marine pollution in Jakarta is quite significant, 700 million rupiah per year was from fishery sector [10]. Moreover, the development of reclamation has also an important impact on the fishery lost up to 207 million rupiah per year. Reclamation also accumulating the additional waste around the seawall in the coastal area that worsen the marine pollution challenges for the community (Fig. 2).

Muara Baru and Kalibaru fishermen villages have been a highly concerned coastal informal area, as they were busy small scale fish industry yet very poor and turning into a continues slum. The rapid economic activities within this area increase the settlement density, most of the informal building construction squeezed each other and turn into a contrast view over the surrounding development. The communities which most of them are small scale fishermen [11] living with the poverty, with the clean water and sanitation issues. With the existing marine pollution around the seawall, the settlement activities also contributing to the accumulation of the stuck waste below the fishermen's elevated houses. This situation was worsened by the neglection attitude of the local communities on their living surroundings. The insufficient land and poor local authority's initiative to provide green open space, waste facilities and management seems added the developing marine pollution issues in the area. As result, the potential of natural coastal area has not been explored for landscape and coastal tourism.

The local fishermen live with poor social economic issues. Most of the children forced to stay with low level of formal education as they need to help the family's economy by involving in the fishing activities [12]. Therefore, the children used to enjoy their daily activities around the dirty and smelly living environment (Fig. 3). Lacking of open and green space gives no alternatives to the local community to have an active social interaction. However, with this poor and unhygienic condition, the local community yet live mostly happily and healthy. The pandemic crisis did not significantly influence their living activities even tough with the economic disruption and adjustment.

Fig. 2 Informal settlement behind the seawall (left) and its' accumulation of marine plastic pollution (right)

Fig. 3 The poor waste management in the fishermen village (left) and children playing around the dirty coastal living environment (right)

The development of informal fishermen village in the Jakarta Northern Coastal area confronts the authorities' regulations on the coastal border, spatial and environmental protection. Many areas are not in accordance with the spatial planning, and some buildings on the river border are classified as illegal and non-permanent buildings. Technically, the regulations have been formulated to reach out to these issues, however it faced some issues including the on-ground conflicts. For instance, the field survey of this study found out that there was the authority's notice board on the government's land ownership installed in the coastline area, yet the community still constructed the permanent houses (Fig. 4).

The major problem such as land legal rights over the uncontrol city population and informal sector settlement development. Indonesian regulations regarding coastal borders have been regulated in Law No. 27, 2007 [13]. Article 31 paragraph 2 stipulates the function of coastal borders as protection against earthquakes, tsunamis, erosion, abrasion, and protecting coastal ecosystems. One of the main objectives of structuring Jakarta's coastal landscape protecting biodiversity, restore green open space for the public and develop ports and docks to support the economy. Spatial regulation for the development of coastal border areas including flats as an alternative to relocate fishermen from the densely populated fishermen villages. Government regulation No. 26, 2007 concerning Spatial Planning Article 31 also mandates the need for provisions regarding the provision and utilization of green open space

Fig. 4 The authority's notice board on the land ownership yet the permanent buildings were densely found in the noticed area

and non-green open space. Green open space clearly functions as lungs, while the non-green open space for community activities.

Changes in the coastline that occur throughout the year as the results of climate change do not really affect the coastal settlement much, because many fishermen's houses were above the water or some of them even living on their boards. The protection and empowerment of fishermen has been regulated in government regulation no. 7 of 2016 [14]. Article 12 has stated the government's protection strategy for fishery businesses, including fishermen. Jakarta spatial planning zoning planning in accordance with statutory regulations regulated the coastal informal settlement to a coastal border. This regulation makes all densely populated settlements in the area need to be relocated. This relocation then followed by the fisheries issue of distanced fishing area to the new set fishermen settlement. Besides, the community also were not used to the new living environment. However, the existing informal coastal settlement still remains un-relocated tough the plan had been widely informed to the communities.

5 Conclusions

Indonesia's coastal area is a richness in natural and spatial resources, economic force and social diversity. Besides the water-based settlement, the coastal communities bring the strong coastal culture. However, the rapid economic and development had caused great environmental degradation in Indonesian coastal area. Marine pollution is seen as an urged coastal problem with complex involvement of causal variables, including the coastal community. This study pinpoints that the fishermen villages take significant role in the causal variables. The marine pollution strongly impacts the living environment of the coastal communities; however, they also take part in contributing to the accumulating waste due to the poor living behavior and insufficient settlement facilities and amenities in the informal setting of fishermen village. The clear, firmed and holistic government legal and regulation on the coastal area as well as its' enforcement is urged to be concerned and formulated.

References

1. RPJMD Jakarta (2018) RPJMD Provinsi DKI Jakarta Tahun 2017–2022. https://bappeda.jakarta.go.id/uploads/document/2018-05-28/63/63__Bab_2_RPJMD_DKI_2022.pdf
2. Rositasari R et al (2017) 5 Dekade LIPI di Teluk Jakarta. Pusat Penelitian Oseanografi-Lembaga Ilmu Pengetahuan Indonesia
3. Coordinating Ministry for Maritime and Investment Affairs of Indonesia (2018)
4. Cordova MR, Nurhati IS (2019) Major sources and monthly variations in the release of land-derived marine debris from the Greater Jakarta area. Indonesia. Sci Rep 9:18730. https://doi.org/10.1038/s41598-019-55065-2
5. Widodo A (2017) Analyzing Indonesia's NCICD project to stop the capital city Sinking. Otoritas: Jurnal Ilmu Pemerintahan 7(2):54–66. Retrieved from https://doi.org/10.26618/ojip.v7i2.769

6. Mardiansyah W (2017) Menelusuri RW 17 Penjaringan, Kawasan Terkumuh di Jakarta. Medcom pada https://www.medcom.id/nasional/metro/0kpnV1qN-menelusuri-rw-17-penjaringankawasan-terkumuh-di-jakarta
7. Loizou E, Chatzitheodoridis F, Polymeros K, Michailidis A, Mattas K (2014) Sustainable development of rural coastal areas: Impacts of a new fisheries policy. Land Use Policy 38:41–47
8. Dsikowitzky L, Ferse SCA, Schwarzbauer J, Vogt TS, Irianto HE (2016) Impacts of megacities on tropical coastal ecosystems—The case of Jakarta, Indonesia. Mar Pollut Bull. (Article in Press)
9. Supartono J, Haluan MFAS, Manuwoto, (2016) Jakarta North Coast development impact on fishery activities. Asian J Sci Res 9:13–23
10. Anna S, Fauzi A (2007) Economi loss of pollution to fisheries: An economic analysis of the Jakarta Bay fisheries. American Fisheries Society Symposium
11. Kusnadi (2002) Konflik Sosial Nelayan: Kemiskinan dan Perebutan Sumber Daya Perikanan. LKIS, Yogyakarta.Mashyuri dan Mochammad N (2000) Pemberdayaan Nelayan Tertinggal: Sebuah Uji Model Penanganan Kemiskinan. Puslitbang Ekonomi dan Pembangunan-LIPI, Jakarta
12. Lwenya C, Yongo E (2021) The fisherman's wife: Vulnerabilities and strategies in the local economy. The Case of Lake Victoria, Kenya. https://doi.org/10.1086/662703
13. Indonesian Government regulation No. 26, 2007
14. Indonesian Government regulation No. 7 of 2016

Urban Development and Traffic Congestion: Jakarta Study during the Pandemic

Garrin Alif Nanditho and Lin Yola

Abstract Rapid economic growth and urbanization of Jakarta has caused socioeconomic challenges in many sectors, including transportation. The city center faces the massive traffic congestion. Urban traffic has contributed the rise on pollutants and energy consumption. For real, this contributes to climate change and global warming. This situation calls for an urge study on the alternative solution on the energy policy to strengthening the current national climate mitigation-related agenda. This study aims to investigate the factors that cause the traffic congestions in order to get alternative solutions to mitigate the impacts on the social, economic and environment. This is a cross and multidiscipline project on urban, spatial, transportation, and policymaking to achieve comprehensive outcomes. The urban traffic in selected major urban routes of city center to buffer zones used the navigation method of PPK GNSS (Post-Processed Kinematic Global Navigation Satellite System) with the aid of Trimble NetR9 receiver as a base station and U-Blox EVK-M8T GNSS low-cost receiver as a rover to get the precise, micro scale and real-time results. This study was conducted during the pandemic, when the traffic volume was not intense. The finding indicated that both macro and micro element of urban development and spatial physical features were the factors of traffic congestion. Overall, this study recommends further studies on the integrated research variables on urban, spatial and environmental as a conceptual Indonesian traffic carbon pricing mode.

Keywords Urban development · Traffic congestion · Jakarta · GNSS · Pandemic

G. A. Nanditho (✉) · L. Yola
Urban Studies Department, School of Strategic and Global Studies, Universitas Indonesia, Jakarta, Indonesia
e-mail: garrin.alif@ui.ac.id

Center for Spatial Data and Analysis, School of Strategic and Global Studies, Universitas Indonesia, Jakarta, Indonesia

E. M. Nia et al. (eds.), *Sustainable Development Approaches*, Lecture Notes in Civil Engineering 243, https://doi.org/10.1007/978-3-030-99979-7_16

1 Introduction

Rapid economic growth and urbanization of Jakarta has caused socioeconomic challenges in many sectors, one of the contributing factors are the transportation sector. The modern urban lifestyle has led to a significant increase in private vehicle ownership. Jakarta is a city with a growth rate of private vehicles reaching 10% annually [1]. However, the continued growth of private vehicles has not been followed by the construction of road infrastructure which only reached 0.1% in the last five years [2]. This indicates that the road infrastructure in DKI Jakarta is still far from ideal, which is one of the factors causing traffic problems, namely congestion. Traffic congestion can cause various bad impacts, especially on the economic sector, such as waste of fuel oil, reduced time productivity, and hampered distribution of goods, and on the environment such as air pollution [3]. The resulting losses reach trillions of rupiah per year. Various policy strategies in controlling traffic congestion have been implemented by the local government, but so far have not been able to provide satisfactory results. That's why it is urgent to study the correlation factors of traffic congestion and the urban variables. This study aims to investigate the factors that cause the traffic congestions in order to get alternative solutions to mitigate the impacts on the social, economic and environment in Jakarta during the pandemic.

2 Literature Review

Urban traffic is the ongoing issue in mostly big cities. One of the urban traffic problems in uncontrolled development cities is the massive traffic congestion. Traffic congestion is a growing problem in urbanizing countries that results in lost time, health problems from pollution, and contributes to the accumulation of greenhouse gas emissions [4]. Studies and methodologies have been explored to get the strategic solutions to this issue. For instance, Dasgupta [5] highlighted the uses of Google Traffic data to measure traffic congestion, with its correlation with other variables such as air pollutions. Study by Noor [6] pointed out that Central Business Districts (CBD) are impeccable areas of traffic attractions exacerbated by the increase of rapid population density, enormous urbanization, and increasing economic and commercial activities. Traffic congestion in the CBD areas is a perpetual problem for sustainable urban development. In recent years, the number of large cities and the incidence of urban haze have been increasing. The expansion of city scale is accompanied by changes in traffic mode and traffic congestion [7]. Identifying congested areas is a complicated task, especially as traffic speed and other census data may not be readily available before a detailed feasibility study is commissioned. Spatial analysis has also been explored in range of methods [8, 9], including for the traffic studies. Methods and GIS (Geographic Information System) is a spatial approach that be used to identify areas with road traffic congestion within cities [10]. There is also an expert system for detecting traffic congestion from real-time GPS data collected from GPS trackers [11]. The GPS technology is a reliable tool in traffic study to get precise data and holistic analysis between the quantitative and qualitative variables. In some context, the duration and time scoping of the study is very essential in investigating the urban traffic. This study uses Post-Processed Kinematic Global Navigation Satellite System (PPK GNSS), as the GNSS technology presents the focus on the precise navigation with the high accuracy [12, 13]. Wu [14] stressed that the traffic state between weekdays and weekends tends to change, the state on weekdays is more regular, whereas on Friday is more complex and random. Therefore, this study explored the range of the urban street traffic variables by using the spatial analysis to get the influencing factors behind the urban congestion problems.

3 Methods

The study area of this paper cover Jakarta major transportation routes (Fig. 1) from city center to buffer zone and buffer zone to city center with the routes of origin–destination; Blok M to Jakarta Kota Station (distance of 15.4 km, with the normal trip duration during the off hours of 46 min) and Jakarta Kota Station to Blok M (distance of 14.3 km with the normal trip duration during the off hours of 58 min). This study used PPK GNSS (Post-Processed Kinematic Global Navigation Satellite System) technology as the data collection method. EVK-M8T U-Blox Evaluation Kit as the rover receiver and Trimble NetR9 as the base station receiver. Private car with the speed of 40–60 kms per hour was used. Meanwhile, the investigation was conducted within in three days field of observation (Monday, Friday, and Saturday), at peak hours (09.00–11.00) and off-peak hours (12.00–14.00). The survey was conducted during the pandemic when the large-scale social restrictions locally known as PSBB (Pembatasan Sosial Berskala Besar) was imposed in Jakarta.

GNSS as the data collection and analysis method included some software; U-Center 20.10, Trimble NetR9 Server 5.22, Convert to RINEX 3.1.4.0, RTKLIB 2.4.2, and RTKLIB 2.4.3. The field survey used the EVK-M8T U-Blox Evaluation Kit as the rover receiver integrated to Laptop with U-Center 20.10 installed. The rover receiver

Fig. 1 The google map showing the location of two case of origin–destination navigated routes

record data from origin–destination routes. The output from rover receiver is 'ubx' data that needed to convert with RTKCONV 2.4.2 to obtain 'obs' data (observation from the rover). Meanwhile, Trimble NetR9 Server 5.22 had been accessed to download 'T02' data to obtain observation from the base station as the correction data and also to download 'P' data as navigation data. Convert to RINEX 3.1.4.0 had been used to convert the 'T02' to get 'o' data (observation from the base station). Then, RTKPOST 2.4.3 had been used to do the PPK (Post Processed Kinematic) method, with 'obs' data as the observation from the rover, 'o' data as the observation from the base station, and 'P' data as the navigation data from the base station to get the high accurate result data. The 'pos' data as the final output can be converted to 'kml' data to be analysed in Google Earth 7.3.4.

4 Results

The result of this study is presented through different form of data that includes duration of trips and the mapping of vehicle movement coordinate distribution. These data indicate the distribution of the traffic within the city area together with the possible causing factors. The post processed kinematic result showed that the survey trip on Monday started from 9.17 to 10.09 for 52 min and 13.43 to 14.24 for 41 min, Friday started from 9.59 to 10.48 for 49 min and 13.46 to 14.36 for 50 min, and Saturday started from 10.16 to 11.01 for 45 min and 13.16 to 13.57 for 41 min (Fig. 2). The result showed the various trip durations over the different routes. Interestingly, many of the trips were recorded taking shorter duration compared to the normal trip duration during the off hours. However, even though Monday morning trip was recorded with longer trip duration due to the busier traffic activities during the first day of the weekdays. The trip duration during weekend (Saturday) was indicated normal with no heavy traffic issue. The study also reported the gap of trip duration among the two routes (Blok M to Jakarta Kota Station and Jakarta Kota Station to Blok M) was indicated not significant. This result shows that the traffic flow and mobility between the direction of the city center to the buffer zone and from the buffer zone to the city center were not showing the big difference. This study reviewed that the stagnant vehicle mobility eased without any traffic problem due to the imposing of the large-scale social restrictions policy by the authority.

The schedule of trip results showed the traffic volume was recorded almost constant during both peak and off-peak hours without a significant gap. The traffic congestion was not a challenge during the pandemic lock down restriction. The traffic volume was heavier during the weekdays as there were still staff activities occur within the office area. However, the traffic was slower in some parts of the roads due to the high volume of cyclist in the left lane during the weekend. There were some roads along the traditional market and the offices blocks were affected by the heavy traffic, the PPK GNSS data showed the coordinate plots (Fig. 3). The traffic was slowing due to the traffics flow from the physical features of multiple roundabouts. However, the signal of the GNSS was not totally distracted due to the

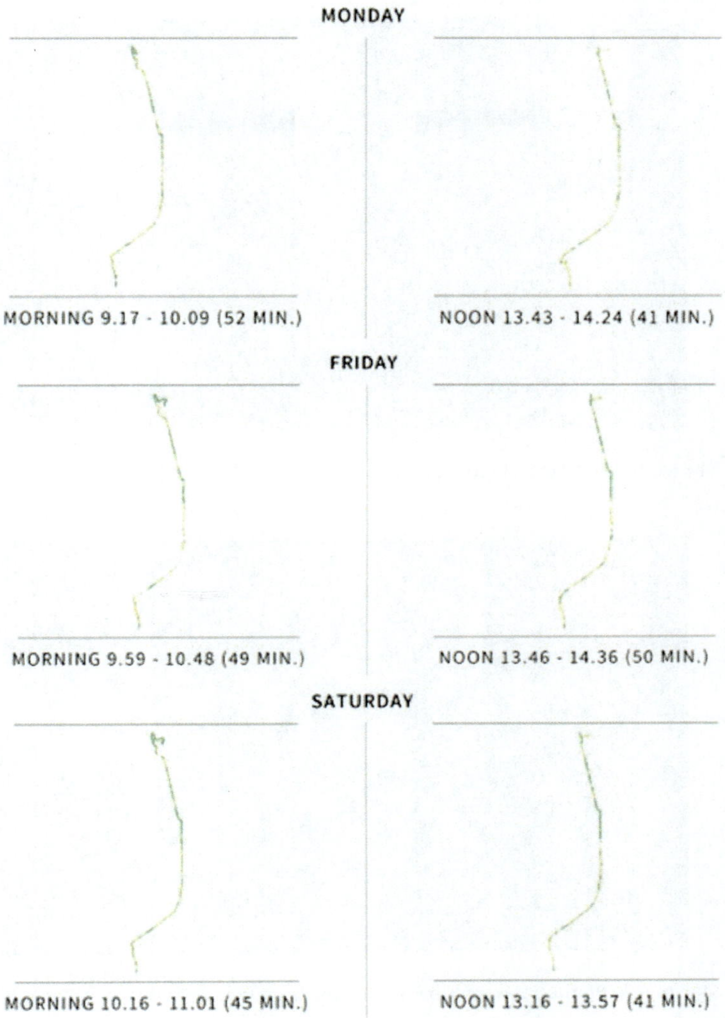

Fig. 2 Weekdays and weekend traffic post processed kinematic navigation

availability of the sky view visible exposure within the underpass (Fig. 4). Figure 5 showed signal error occurred within the high-rise buildings' canyon within the underpass tunnel. It indicated that the vertical and horizontal surface distraction affected the signal.

The traffic policies to control the traffic in the main routes in the city center of Jakarta were not applied due to the pandemic large-scale social restrictions policies, therefore, it didn't affect the traffic significantly compared to the normal days before the pandemic when the authority imposed the even and odd vehicle plat number shifting. However, this study recorded the slowing down traffic around the traffic light area in the city center even tough it did not cause the traffic jam (Fig. 6).

Fig. 3 The traffic along the office area

Fig. 4 Slowing-down traffic in multiple roundabouts and underpass

Contrary, the traffic was easing well in some parts of the roads along the city center towards the remote area and the other way round. The signal was also recorded smoothly. Therefore, besides the impact of the traffic policy on the traffic flow, this study analyzed that the physical features (e.g. roundabouts, high-rise buildings, trees and tunnels) and the city land use and landmarks (e.g. institutional, commercial and city amenities) also contributes to the traffic in both city center and buffer zones.

Fig. 5 Signal error in between high-rise building canyon within the underpass tunnel

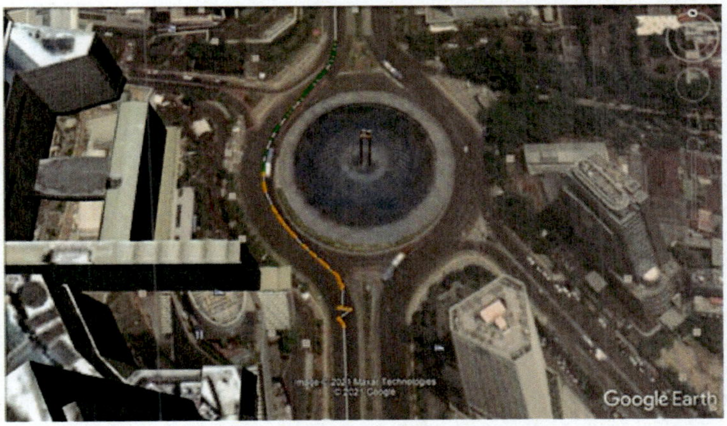

Fig. 6 Slowing down traffic within the traffic light area

5 Conclusions

The conclusions of this study highlights that the distribution of traffic was various in both city center and buffer zone, the traffic conjunction factor includes the landuse location and the road sizes. The traffic volume was not too high due to the imposing of the pandemic social gathering limitation policy by the authority. There was no significant gap recorded between the weekdays and the weekend traffic. There was signal error found in some coordinates due to the physical obstruction, such as tunnels, trees, and high-rise building. The navigation method of PPK GNSS (Post-Processed Kinematic Global Navigation Satellite System) with the aid of Trimble NetR9 receiver as a base station and U-Blox EVK-M8T GNSS was reviewed as a reliable analysis tool for this study. The finding of this study is recommended to consider in making better traffic and pandemic social gathering limitation policy. This study also pinpoints that the poor land use planning and city development greatly contributes to the city traffic congestion problem. However, due to the limitation of research scope and investigation time and resources, this study suggests further analysis and studies with more variables such as routes, times, type of vehicles, and environmental features for a more holistic consideration and findings.

References

1. Raharjo DB, Fakhri FM (2020) Tingkat Kemacetan Tak Berkurang, Pemprov DKI: Capaian Luar Biasa. Available at: https://www.suara.com/news/2020/02/04/125806/tingkat-kemacetan-tak-berkurang-pemprov-dki-capaian-luar-biasa?page=1 (28 September, 2021)
2. Samudra, MA (2019) Pantas Sering Macet, Panjang Jalan di Jakarta Cuma Segini, Ternyata Ini Idealnya. Available at: https://www.gridoto.com/read/221875387/pantas-sering-macet-panjang-jalan-di-jakarta-cuma-segini-ternyata-ini-idealnya (28 September, 2021)
3. Mirlanda AM (2011) Kerugian Ekonomi Akibat Kemacetan Lalu Lintas di Ibukota. Universitas Indonesia, Depok
4. Brent D, Louis-Phillipe B (2020) Traffic congestion, transportation policies, and the performance of first responders. J Environ Econ Manage 103:102339
5. Dasgupta S, Somik L, David W (2021) Spatiotemporal analysis of traffic congestion, air pollution, and exposure vulnerability in Tanzania. Sci Total Environ 778:147114
6. Noor MA., Ashrafi S, Fattah MA, Morshed SR, Rahman S (2021) Assessment of traffic congestion scenario at the CBD areas in a developing city: In the context of Khulna City, Bangladesh. Transp Res Interdisc Perspect 11:100435
7. Lu J, Bin L, He L, Abdo A (2021) Expansion of city scale, traffic modes, traffic congestion, and air pollution. Cities, The Int J Urban Policy Plann 108:102974
8. Yola L, Siong HC (2017) Computer simulation as an alternative approach in climatically responsive urban configuration study. Chem Eng Trans 56:505–510. https://doi.org/10.3303/CET1756085
9. Yola L, Ayegbusi OG, Saiya HG, Djaja K (2022) Empirical and numerical approaches in urban microclimate modeling: Investigation on the reliability. Lecture Notes in Civil Engineering 2022(161):51–57. https://doi.org/10.1007/978-981-16-2329-5_7
10. Loo BPY, Zhiran H (2021) Delienating traffic congestion zones in cities: An effective approach based on GIS. J Transport Geogr 94:103108
11. D'Andrea E, Marcelloni F (2017) Detection of traffic congestion and incidents from GPS trace analysis. Expert Syst Appl 73:43–56
12. Paziewski J, Crespi M (2020) High-precision multi-constellation GNSS: Methods, selected applications and challenges. Meas Sci Technol 31(1)
13. Martin S, Alastair P (2021) Accuracy and precision of GNSS in the field. GPS and GNSS Technology in Geosciences, pp 393–414
14. Wu J, Xubing Z, Yi P, Xiaojun Z (2021) Recurrence analysis of urban traffic congestion index on multi-scale. Physica A: Stat Mech Its Appl 585:126439

Influence of Urban Canyon Direction on Long Wave Radiation Pattern in a Tropical Context

Lin Yola

Abstract Poor urban microclimate and thermal comfort has contributed a continues urban energy issue. The urban energy imbalance is caused by the uncontrolled urban development and ignorance to the urban environmental damages. The configuration of urban spaces and development play significant role in the level of urban thermal heat. The high intensity of solar radiation in tropical region makes the urban thermal discomfort a major challenge in creating a comfortable and lively outdoor urban space. The lacking of firmed studies on the relationship of urban configuration and microclimate were the initial background of this study. This study aims to investigate the long wave radiation behavior in four scenario of urban canyon directions. This study was situated in high density residential area of Kuala Lumpur, Malaysia using the ENVI-met V 3.1 simulation. The finding indicated that the East–West direction received inconsistent pattern and significant intensity of long wave radiation over the South-North. This study suggests the strategized and the passive design of urban form to be considered for achieving the climatically responsive urban configuration in the tropical context.

Keywords Urban Canyon · Long wave radiation · Tropical context · ENVI-met · Kuala Lumpur

1 Introduction

Climate change is a global threat that impacts almost all sectors, including the urban life. Therefore, the climate change mitigation and adaptation agenda are still focusing on the city as central position of the agenda [1]. This situation is due to the rapid economic growth and urban development that brings the complex social, political and economic issues. The poor urban microclimates and thermal discomfort causes the phenomena of Urban Heat Island (UHI), the major climate change contributor from urban sector. Both urban microclimate and thermal comfort were highlighted

L. Yola (✉)
Department of Urban Studies, School of Strategic and Global Studies, Universitas Indonesia, Jakarta, Indonesia
e-mail: lin.yola@ui.ac.id

E. M. Nia et al. (eds.), *Sustainable Development Approaches*, Lecture Notes in Civil Engineering 243, https://doi.org/10.1007/978-3-030-99979-7_17

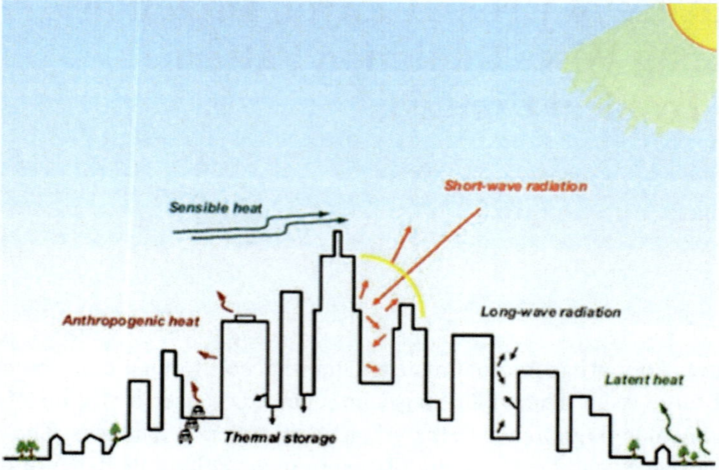

Fig. 1 The heat components in the urban energy balance system. *Source* US Environmental Protection Agency (2013)

significantly influenced by the configuration of urban fabric. The ongoing discussion on the correlation between the urban configuration on the urban microclimates and thermal comfort has been explored in the urban energy balance studies. The poor urban configuration settings have caused the modification of heat components that makes the unsustainable urban energy balance (Fig. 1).

Short wave and long wave radiation are two main heat variables that contribute to the urban energy balance. In a high urban density area, the long wave radiation traps and stores the heat in the urban surface and contributes to the high intensity of Urban Heat Island (UHI) in tropical area. Besides the solar radiation, urban wind also plays significant role in the modification of urban energy balance system, as it mainly eases the heat from one to another urban spaces. Building position and the role of canyon were among the strategies highlighted to maximise the urban wind circulation as well as to minimise the risk of heat absorption [2–8]. Longwave radiation is strongly determined by the geometry of the surface when it absorbs and traps the radiation heat. The high long wave radiation mainly occurs in the dense cities with the low albedo which leads to poor Urban Heat Island (UHI). As the solar radiation set the main source of heat to the modification of microclimate and thermal comfort within the urban space, the need of focus study on the solar radiation especially the long wave radiation is an urgent stage for achieving the sustainable alternative of urban configuration strategies. Therefore, this study investigated the impact of choice of urban configuration with the different setting of canyon features on the intensity of long wave radiation.

2 Literature Review

In the pioneer urban energy studies, solar radiation was discussed as the main source of heat in urban energy balance [9, 10]. The reflection and absorption of short and long wave solar radiation (Fig. 2) results in the modification of main climate variables such as surface temperature, air temperature, mean radiant temperature, and thermal comfort. The micro-scale physical features between buildings in urban spaces cause major heat generators [11]. The study on strong relation of urban climate and urban configuration has been widely explored. Some of studies had discover the urban strategies to modify the solar radiation; including urban fabric and geometry [12–14], urban space structure [15] and building shape and orientation [16, 17].

As the short-wave radiation is emitted directly from the solar radiation and surface while the atmosphere discharges the long wave radiation [10]. Therefore, the urban outdoor heat stress would be influenced by short wave radiation during the daytime and long wave radiation during the night-time. The direct radiation and diffuse radiation are mostly influenced by the vertical obstruction on the ground surface [18, 19]. Pioneer study [3] formulated the Height to Width (H/W) aspect ratio and Sky View Factor (SVF) on the canyon configuration and its' impact on the maximum heat island stress. It was stressed that the maximum air temperature exposed to the canyon space were for highest H/W aspect ratio and smallest SVF [20]. In this context, both H/W and SVF determines the intensity of the affecting long wave radiation within the canyon space. However, the direction of canyon was not detailly discussed in the said formula [21, 23]. Therefore, this study aimed to investigate the impact of

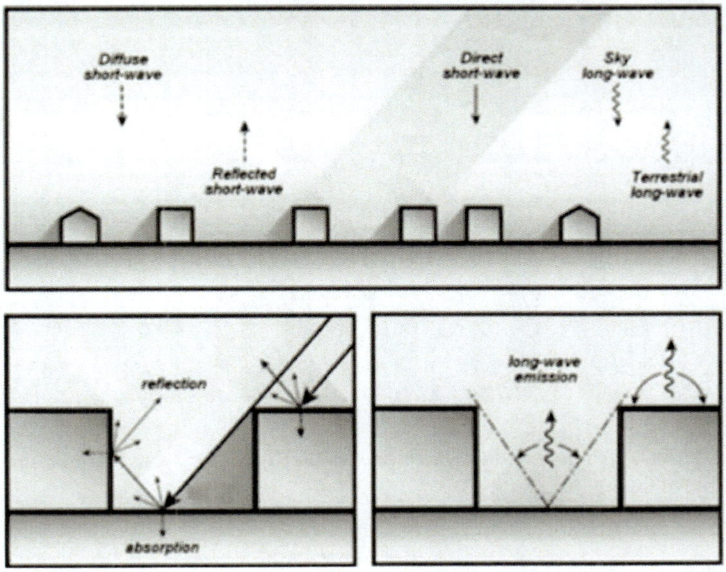

Fig. 2 Solar radiation short wave and long wave on the urban surface. *Source* Erell et al. [10]

urban configuration alternatives on the intensity of long wave radiation in the tropical context of Kuala Lumpur.

3 Methods

Two empirical sites of residential blocks in Kuala Lumpur city were investigated in this study, with the urban configuration scenario with the canyon direction of East – West (parallel with sun path) and North–South (perpendicular to the sun path). Both sites are with the high-rise residential blocks in Kuala Lumpur. The existing site urban configuration is Courtyard Canyon with a canyon in the center of the outdoor courtyard. This study set other three urban configurations for simulation; Courtyard, U and Canyon. The four urban configurations were investigated with the said two scenarios of East–West and North–South (Fig. 3).

ENVI-met V3.1 Beta simulation used in this study, with the grid cells of x = 210, y = 210 and z = 30. The site coordinate was set in Kuala Lumpur, Malaysia (3O8'51"N101O41'36"E). The model was simulated for 21 June 2015 with the simulation time of 24 h. The microclimatic data were inserted as follow: initial temperature of 303.15 (°K), specific humidity of 4, relative humidity of 83%, wind speed of 1.4 m/s, wind direction of 225 (South West), albedo walls of 0.3 and albedo roofs of 0.5.

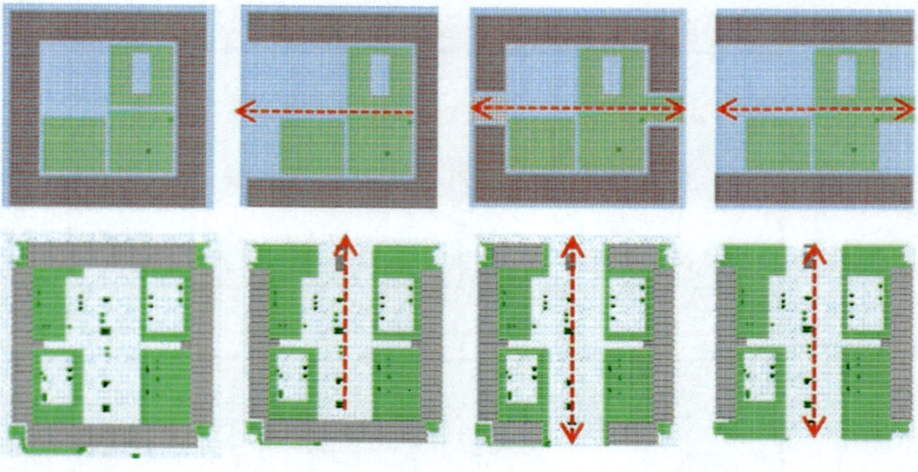

Fig. 3 The four urban configurations setting in ENVI-met for two Canyon direction scenarios: East–West and South–North

Table 1 Mean of long wave radiation (W/m^2) in two scenarios of Canyon direction

Urban configurations	Canyon direction facing East–West	Canyon direction facing South–North
Courtyard	488.60	518.39
U	489.27	514.57
Courtyard Canyon	466.38	513.44
Canyon	472.18	509.64

4 Results

This study presents two sets of 24 h records of long wave radiation according to the setting canyon directions; East–West and South North (Table 1). The results show that there was significant gap between the diurnal and nocturnal long wave radiation pattern. Overall, the mean of long wave solar radiation was recorded gapping between the two settings of canyon direction East–West and South-North. Data show that the urban configurations with canyon direction facing East–West (Table 2) cause lower long wave radiation over the South-North (Table 3). This situation indicated positive correlation with the earlier discussion that the trapped and absorbed heat within the enclosed area results in high long wave radiation intensity, especially during the night.

Tough the solar radiation was fully received throughout the day; the long wave radiation was recorded high during the night time. The significant gap was also found in the long wave radiation pattern in four urban configurations when canyon set facing East–West. However, the uniform pattern was recorded when canyon feature was directed to South–North. Technically, the urban configuration with canyon features in East–West scenario (occurs in Canyon and Courtyard Canyon configuration) results in low long wave radiation intensity over the urban configurations without canyon features (Courtyard and U configuration). In this tropical context, the recommended urban configuration with high to low SVF that set better scenario of long wave radiation are; Canyon, Courtyard Canyon, U and Courtyard.

Table 2 Diurnal and Nocturnal long wave radiation with Canyon facing East–West

Diurnal long wave radiation (W/m^2)

Urban configuration	7am	8am	9am	10am	11am	12pm	1pm	2pm	3pm	4pm	5pm	6pm	Mean
Courtyard	475.94	476.55	480.88	484.81	488.31	492.28	502.64	508.58	508.74	513.96	500.73	496.21	494.14
U	473.92	473.81	477.82	483.79	488.68	492.61	505.64	509.52	510.35	508.99	504.75	504.67	494.55
Courtyard Canyon	455.5	453.63	467.29	480.69	484.84	469.31	477.2	501.94	503.24	500.88	480.95	473.7	479.10
Canyon	462.41	460.39	475.13	490.8	500.56	505.24	507.99	508.34	506.96	503.09	494.93	483.02	491.57

Nocturnal long wave radiation (W/m^2)

Urban configuration	7pm	8pm	9pm	10pm	11pm	12am	1am	2am	3am	4am	5am	6am	Mean
Courtyard	491.79	489.2	487.28	485.64	484.2	482.91	481.74	480.67	479.67	478.73	477.86	477.04	483.06
U	494.69	491.68	489.25	487.1	485.23	483.62	482.21	480.96	479.82	478.76	477.76	476.81	483.99
Courtyard Canyon	463.94	459.1	456.69	455.03	453.76	452.72	451.87	451.16	450.58	450.09	449.69	449.38	453.67
Canyon	466.5	460.2	457.19	455.08	453.37	451.93	450.67	449.56	448.55	447.62	446.76	445.96	452.78

Table 3 Diurnal and Nocturnal long wave radiation with Canyon facing South-North

Diurnal long wave radiation (W/m^2)

Urban configuration	7am	8am	9am	10am	11am	12pm	1pm	2pm	3pm	4pm	5pm	6pm	Mean
Courtyard	476.61	476.31	484.54	521.53	558.36	584.8	602.64	568.68	546.97	578.92	568.84	531.98	541.68
U	476.35	475.83	482.67	517.97	552.94	578.38	595.57	562.18	541.22	572.6	562.5	526.59	537.07
Courtyard Canyon	475.85	475.06	481.87	518.04	553.71	578.87	595.72	562.36	541.69	572.71	562.39	526.67	537.08
Canyon	475.18	473.94	479.61	515.18	550.07	574.79	591.27	558.44	538.54	568.98	557.89	522.61	533.88

Nocturnal long wave radiation (W/m^2)

Urban configuration	7pm	8pm	9pm	10pm	11pm	12am	1am	2am	3am	4am	5am	6am	Mean
Courtyard	516.17	508.04	503.06	499.38	496.43	493.94	491.77	489.82	488.04	486.37	484.81	483.32	495.10
U	511.93	504.38	499.71	496.24	493.44	491.07	488.99	487.12	485.38	483.76	482.22	480.75	492.08
Courtyard Canyon	511.64	503.45	498.29	494.45	491.36	488.74	486.42	484.33	482.39	480.57	478.84	477.2	489.81
Canyon	507.82	499.36	494.04	490.1	486.94	484.26	481.91	479.78	477.82	475.98	474.24	472.59	485.40

5 Conclusions

The setting of urban configuration plays significant impact on the pattern of long wave radiation modification. Tough this study did not investigate the correlation with the modification on other microclimate features, the intensity of long wave radiation was found varies in different altitude time. The orientation of canyon feature towards the East and West mainly influenced the intensity of long wave radiation, with the significant gap between the diurnal and nocturnal. In this context of study, the urban configuration with canyon feature facing the East–West recorded a more variety of long wave compared to the South-North. One of the contextual main factors was the wind direction facing South-West. However, urban configuration with canyon feature (mainly canyon and courtyard canyon) performed lesser heat stress over a non-canyon urban configuration. The findings pinpointed those alternative strategies of canyon direction in urban configuration creates significant impact to the outdoor heat stress of tropical urban environment.

References

1. Carter JG, Cavan G, Connelly A, Guy S, Handley J, Kazmierczak A (2015) Climate change and the City: Building capacity for urban adaptation. Prog Plan 95:1–66
2. Nunez M, Oke TR (1977) The energy balance of an urban Canyon. J Appl Meteorol 16(1):11–19
3. Oke TR (1987) Boundary Layer climates. Routledge, New York
4. Oke TR (1988) Street design and urban Canopy Layer climate. Energy and Buildings. 11:103–113
5. Elhanas MM (2003) The effects of urban configuration on urban air temperatures. Archit Sci Rev 46(2):2003
6. Emmanuel MR, Rosenlund H, Johansson E (2007) Urban shading—A design option for the tropics? A study in Colombo, Sri Lanka. Int J Climatol 27(14):1995–2004
7. Emmanuel MR (2005) An urban approach to climate-sensitive design: Strategies for the tropics. Spon Press, New York
8. Lim J, Ooka R (2014) Building arrangement optimization for urban ventilation potential using genetic algorithm and CFD simulation. The 31st International Symposium on Automation and Robotics in Construction and Mining (ISARC 2014)
9. Gartland LM (2008) Heat Islands: Understanding and mitigating heat in urban areas. Earthscan
10. Erell E, Pearlmutter D, Wiliamson T (2011) Urban microclimate: Designing the spaces between buildings. Earthscan
11. Weber F (2015) Towards mitigative buildings and urban environments Tthe 31th International PLEA Conference Passive Low Energy Architecture. 9–11 September. Bologna
12. Martin L, March L (1972) Urban space and structures. Univesity Pers, Cambridge
13. Shashua-Bar L, Tzair Y, Hoffman ME (2004) Thermal effects of building geometry and spacing on the urban Canopy Layer microclimate In A hot-humid climate In summer. Int J of Climatol 24:1729–1742
14. Johansson E (2006) Influence of urban geometry on outdoor thermal comfort in a hot dry climate: A study in Fez. Morocco. Build Environ 41(2006):1326–1338
15. Hagen K, Gasienica-Wawrytko B, Loibl W, Pauleit S, Stiles R, Totzer T, Timmel H, Kostl M, Feilmayr W (2014) Smart environment for smart cities: Assessing urban fabric types and microclimate responses for improved urban living conditions. REAL CORP 2014 Tagungsband. 21–23 May 2014,Vienna, Austria

16. Gerber DJ, Lin S (2014) Geometric complexity and energy simulation. Proceedings of the 18th International Conference on Computer-Aided Architectural Design Research in Asia (CAADRIA 2013), pp 87–96
17. Ling CS, Ahmad MH, Ossen DR (2007) The effect of geometric shape and building orientation on minimising solar insolation on high-rise buildings in hot humid climate. J Constr Devel Countries 12 (1)
18. Givoni B (1998) Climate considerations in buildings and urban design. Van Nostrand Reinhold
19. Pearlmutter D, Berliner P, Shaviv E (2006) Physical modeling of Pedestrian energy exchange within the Urban Canopy. Build Environ 41(2006):783–795
20. Yola L, Ho CS (2017) Computer simulation as an alternative approach in climatically responsive urban configuration study. Chem Eng Trans 56(2017):505–510
21. Yola L. (2020). Canyon effects in urban configurations: Tropical context study. IOP Conference Series: Earth and Environmental Science 436(1), art. no. 012028. https://doi.org/10.1088/1755-1315/436/1/012028
22. Yola L, Siong HC, Djaja K (2020). Climatically responsive urban configuration in residential area: Research gaps. AIP Conference Proceedings, 2255, art. no. 0700141. http://scitation.aip.org/content/aip/proceeding/aipcp, ISBN: 978–073542014–4. https://doi.org/10.1063/5.0013796

Walkabilty Study on Pedestrian Path in the Rawamangun Velodrome Area, East Jakarta

Bachtiar Marpaung and Lin Yola

Abstract Pedestrian paths in the Rawamangun Velodrome area have been built since the start of the 2018 Asian Games, but the results of observations of the use of pedestrian paths are still low in user traffic. This writing aims to identify the walkability index and pedestrian facilities based on pedestrian service standards. This study used the mixed method for investigating the pedestrian path facilities, pedestrians volume by and the width of the pedestrian path in each observation segment. The walkability index was measured by using the formula and theory of the global walkability index. The value of the global pedestrian index analysis was recorded as 57.63, which means that some facilities around the pedestrian path (sidewalk) can be reached or passed on foot, but the fact is that there are frequent conflicts between pedestrians and motorcycle parking lots and street vendors. The finding of this study could be used for policy evaluation of the walkable area of Velodrome Rawamangun.

1 Introduction

The ever-increasing population growth will affect the city's transportation facilities and infrastructure. This causes transportation facilities to be able to serve all components of road users from four-wheeled vehicles such as cars, two-wheelers such as motorcycles and bicycles as well as pedestrians [1–4]. Walking is the main mode of transportation for most people, especially metropolitans in the world [2, 5, 6]. For a big city like DKI Jakarta that experiences daily traffic jams due to the lacking of efficient transportation management and policies, this problem results in an increased risk of environmental, economic and health issues. The revitalization of pedestrian paths or sidewalk has begun to be improved since the 2016 Fiscal year. The DKI Jakarta Regional Government since 2016 has had a pedestrian path revitalization

B. Marpaung (✉) · L. Yola
Department of Urban Studies, School of Strategic and Global Studies, University of Indonesia, Depok, Indonesia
e-mail: bachtiar.marpaung@ui.ac.id

L. Yola
e-mail: lin.yola@ui.ac.id

E. M. Nia et al. (eds.), *Sustainable Development Approaches*, Lecture Notes in Civil Engineering 243, https://doi.org/10.1007/978-3-030-99979-7_18

program with a target of 134 km sidewalks from 2017 to 2019. It is hoped that the sidewalk construction to integrate public transportation in the city of Jakarta is in line with the Governor's Instruction Number 66 of 2019 [7] regarding Jakarta's air quality control. Pedestrian users in Jakarta were encouraged to do more walking activities and use public transportation more often, in work activities, or in sports.

This study investigates the Rawamangun Velodrome area of Pulogadung sub-district in East Jakarta as the study area. This area consists of commercial areas, office areas, settlements, and leisure and sports areas. Priority has been given to the construction and construction of pedestrian path from the fiscal year of 2016 to 2018 due to the serving the purpose of hosting the 2018 Asian Games, as this area provides the bicycle racing division.

This paper aims to evaluate and analyze the walkability index of the study area, by using the Global Walkability Index (GWI) parameters, and whether the study area met pedestrian service standards in accordance with the Regulation of the Minister of Public Works No. 3/PRT/M/2014 dated 26 February 2014, concerning Planning, Provision and Utilization of Pedestrian Network Infrastructure and Facilities in Urban Areas.

2 Literature Review

Walkability is a comprehensive assessment of all aspects of the activity and walking path environment. Walkability can also be used to reflect the conditions and feasibility of walking in an area. The basic theory of walkability assessment procedures can provide an overview and assess connectivity or connectedness, the quality of sidewalks, footpaths, in cities [8–11]. Various methods for assessing the walkability index have been developed in ranges of developed and developing country. The walkability assessment method is generally carried out using two study approaches, namely subjective and objective [9–13]. The available data can be further classified into qualitative and quantitative data.

However, the walkability assessment methods that have been developed in various developed countries cannot be simply applied in Indonesia because these methods are developed based on conditions and needs that are not necessarily appropriate to the situation in Jakarta or Indonesia. This has an impact on the level of reliability and relevance. Therefore, of the various methods currently available, only the method that uses an objective approach and combines quantitative–qualitative data that can be applied in the assessment of walkability in urban areas of Indonesia.

3 Methods

The research method of the Global Walkability Index, conducted and improvised by Holly Virginia Kreambeck for the World Bank [9], describes a qualitative assessment and analysis of the condition, environmental comfort of pedestrian paths, their safety and security [10].

This research will conduct a walkability assessment. The method of obtaining the walkability index has eleven parameters. Parameter observations will be carried out by assessing pedestrian conflicts with motorized vehicles or cyclists, parameters of availability and cleanliness of pedestrian paths, availability and safety of crossings, parameters of motorized vehicle behavior, availability of supporting facilities for groups of people with disabilities, security parameters from criminal acts.

To obtain data and facts in the field, the author will use the observation method, by taking photos of the observation area and also taking notes to get a general picture of the observation area. Assessment is carried out with a score of 1 to 5 (1 is the lowest score and 5 is the highest score).

This method includes the number of pedestrians from the calculation of the pedestrian flow for 15 min, as well as in this research, because the authors want to know the number of sidewalk users in a matter of every fifteen minutes in one observation.

The Ministry of Public Works has issued a guide to the planning procedures for the development of pedestrian facilities, which can be used as a reference for planning and evaluating sidewalk facilities, especially in urban areas in Indonesia [14, 15]. The pedestrian service standards contained in these guidelines are technical and universal and adapt to existing environmental conditions. Standards for sidewalk space can be formulated and used according to the types of pedestrian sections and considering local habits and types of activities.

After all the data is collected, then it is calculated and analyzed to get the value of the walkability index as developed by the results of previous studies. For ease of calculation, the value of the assessment score is converted in the range of values 0–100. The assessment is carried out with a score of 0–24 to 90–100 (0–24 is the lowest score, and the 90–100 is the highest score).

After calculating the walkability index, the pedestrian flow was calculated based on the total number of pedestrians passing through the observed segment using the technical guidelines of the Ministry of PUPR (Minister of Public Works 2014). Observations were made during peak hours with 15-min intervals. The results are compiled every 15-min intervals and the total number of pedestrians is calculated and adjusted into units, with the following formula: Pedestrian Flow = ((pedestrians/m)/minute).

4 Results Discussion

The results of the overall parameters and observation points produce an average value of 57.63 which means that according to Table 1, several facilities around the pedestrian path (sidewalk) can be reached or passed on foot. The walkability assessment per parameter for each observation point and segment is shown in Table 1.

The highest walkability value for all segments is obtained in parameter six, namely supporting facilities or amenities (62.70), whereas the lowest value is in parameter

Table 1 The results of the calculation of walkability index at 9 observation points

No	Parameter assessment	Observation point									WI Score
		I	**II**	**III**	**IV**	**V**	**VI**	**VII**	**VIII**	**IX**	
1	Pedestrian Conflict with Other Transportation	4	4	3	4	5	4	4	5	5	60.00
2	Availability of Pedestrian Paths (With Maintenance And Cleaning)	4	4.5	3.5	4	5	4	4	5	5	60.66
3	Crossing Availability	4	4	3	3	4	3	3	4	4	51.13
4	Crossing Safety	4	4	3.5	3.5	4	2	4	4	4	52.42
5	Motorcyclist Behavior	4	4	4	4	4	3	4	5	5	59.51
6	Supporting Facilities (Amenities)	5	5	4	3	4	3	4	5	5	62.70
7	Infrastructure for People with Disabilities	4	4	3	3	5	3	3	5	5	57.66
8	Barrier or Obstacle	4.5	4	3.5	3.5	5	3.5	4	5	5	61.21
9	Security From Crime	4	5	4	3	4	3	4	4	4	53.43
10	Number of Pedestrians (within 15 min)	57	38	36	43	36	33	38	62	72	
11	Length of Pedestrians observation	0.52	0.2	0.1	0.2	0.1	0.1	0.2	0.2	0.52	
The average of WI Score											**57.63**

three, namely the availability of crossings (51.13). Related to the results of the walkability index, that the results of observations for all segments and observation points score between 4 and 5, there is indeed a score of 3, but this is an assessment in certain segments, supporting facilities available; blind tile guides, park benches, motor barrier poles (bollard), night lights, trash cans, and the availability of reforestation lanes, as well as some segments providing manholes (needs for installing PLN utilities, PAM, internet cables.) PJU lights), and the average sidewalk width is above 2.5 m' according to the 2014 PUPR Ministerial Regulation as a minimum standard of 2 m' for residential areas, offices, shops, recreation, bus terminals and schools.

As the lowest score for the availability of crossing parameters, because in field observations, zebra cross signs are only found at red light intersections, while the farthest distance for zebra crossings is an average of 200 m' which causes many pedestrians to cross spontaneously or carelessly. Actually, this does not need to happen, because technically the placement of a zebra cross or crossing sign and bus shelter facilities have been placed according to the 2014 PUPR Ministerial Regulation which is a minimum of 300 m', it is up to road users whether they are aware of safety in crossing or ignore it.

5 Conclusions

This study highlights that the walkability index of the Velodrome Rawamangun was indicated as the average level of the road section, the total walkability index at the observation point was reported as 57.63. The results show that several facilities around the pedestrian path of the Velodrome Rawamangun can be reached and traversed. This study also pinpoints the availability of supporting facilities (amenities) was seen as the strength of the area, which claimed by the DKI Regional Government as pedestrians 'friendly' place by providing blind guide tiles, bollards, park benches, streetlights, bus shelters and greening lanes. Meanwhile, for the criteria of the service standard, was recorded as the standard A or the pedestrian flow was not too congested, only <6.7 people/meter/minute.

The assessment of the walkability index in this study limits the study on the overview of the quality of pedestrian facilities in and around the Rawamangun Velodrome area. The expected improvement is suggested to increase the number and awareness of pedestrians using lanes to reach each desired facility, without being destructed by external factors such as crowds of drivers and parking lots, or illegal hawkers especially during off hours. Pedestrians and other users expected to socially utilize the spaces to maximize the use of sidewalks and the walkable environment.

References

1. Adams MA, Todd M, Kurka J, Conway TL, Cain KL, Frank LD, et al (2015) Patterns of walkability, transit, and recreation environment for physical activity. Am J Prev Med [Internet] 49(6):878–87. Available from: https://www.scopus.com/inward/record.uri?partnerID=HzOxMe3b&scp=84949725044&origin=inward
2. Golan Y, Henderson J, Wilkinson NL, ... (2019) Gendered walkability. J Transp ... [Internet]. Available from: https://www.jstor.org/stable/26911279
3. Tanan N, Wibowo SS, Tinumbia N (2017) Pengukuran Walkability Index pada Ruas Jalan di Kawasan Perkotaan (Walkability Index Measurement on Road Links in Urban Area). J Jalan-Jembatan [Internet]. Available from: http://jurnal.pusjatan.pu.go.id/index.php/jurnaljalanjembatan/article/view/90
4. Collins PA, Tait J, Fein A, Dunn JR (2018). Residential moves, neighbourhood walkability, and physical activity: A longitudinal pilot study in Ontario Canada. BMC Public Health [Internet] 18(1). Available from: https://www.scopus.com/inward/record.uri?partnerID=HzOxMe3b&scp=85050855073&origin=inward
5. Tanan N, Darmoyono L (2017) Achieving walkable city in Indonesia: Policy and responsive design through public participation [Internet] 1903, AIP Conference Proceedings. Available from: https://www.scopus.com/inward/record.uri?partnerID=HzOxMe3b&scp=85035238619&origin=inward
6. Lo RH (2009) Walkability: What is it? J Urban 2(2):145–166
7. Governor of the Special Capital Region of Jakarta Province (2019) Instruksi Gubernur Provinsi Daerah Khusus Ibukota Jakarta Nomor 66 Tahun 2019 047:583–606
8. Forsyth A (2015) What is a walkable place? The walkability debate in urban design. URBAN Des Int [Internet] 20(4):274–92. Available from: https://doi.org/10.1057/udi.2015.22
9. Krambeck H, Shah JJ (2006) The global walkability index
10. Leather J, Fabian H, Gota S, Mejia A (2011) Walkability and Pedestrian facilities in Asian Cities State and issues. Asian Dev Bank Sustain Dev Work Pap Ser. 17:69
11. Nyagah P (2015) A multi-procedural approach to evaluating walkability and Pedestrian safety [Internet]. Available from: http://digitalscholarship.unlv.edu/thesesdissertations/2568
12. Alves F, Cruz S, Ribeiro A, Silva AB, Martins J, ... (2020) Walkability index for elderly health: A proposal. Sustain [Internet]. Available from: https://www.mdpi.com/2071-1050/12/18/7360
13. Cambra P (2020) How does walkability change relate to walking behavior change? Effects of a street improvement in pedestrian volumes and walking experience. J Transp Heal [Internet] 16. Available from: https://www.scopus.com/inward/record.uri?partnerID=HzOxMe3b&scp=85075400871&origin=inward
14. Menteri Pekerjaan Umum (2013, 2014) Pedoman Perencanaan, Penyediaan, dan Pemanfaatan Prasarana dan Sarana Jaringan Pejalan Kaki di Kawasan Perkotaan. Menteri Pekerj Umum Republik Indones [Internet]:8. Available from: http://pug-pupr.pu.go.id/_uploads/Produk_Pengaturan/Permen%20PUPR%20No%2003-2014.pdf
15. Kementerian Pekerjaan Umum dan Perumahan Rakyat (2017) Pedoman Bahan Konstruksi Bangunan dan Rekayasa Sipil: Perencanaan Teknis Fasilitas Pejalan Kaki. SE Menteri PUPR:5–6

LRT's Transit Oriented Development in Jakarta, Bogor and Bekasi: The *Urban Infrastructure Development vs Economic Interest*

Oktia Hendra and Lita Sari Barus

Abstract What is so special about Jakarta? It is the capital city of the Republic of Indonesia, and just as central place theory of Christaller, it was becoming the heart of this country. Almost 30 percent of Indonesia GDP is derived from this area. So does the chains of command where executives, legislatives and supreme court based in this "Special" City. Consequently, Jakarta has become magnets for urban dreamers. For centuries people from all over Indonesia migrates to this city for its fortune; either from trading and merchandising, Political Career, down to workers on public and private sectors. Amazingly as Central of Government as well as central for economy, Jakarta still exists despite of huge city problems and complicated urban structures. Along with Jakarta's development, the price for property is rising exponentially for the last 50 years. Price of a square meter land in Menteng Jakarta for example, might exceed the price of a square meter land in Malibu-California. The expansion of infrastructure to the surrounding area of Jakarta at the late of last century create circumstances where demands for rural area development also rose. Suburban area became choices for cheaper housing demands. From 1990 up to now, the growth of suburbs has increased exponentially. The area widely spread from Banten, Tangerang, Depok, Bekasi Further to Karawang. In Northern Part of Jakarta, some high-end developers even reclaimed the shore and creates new elite's islands for the rich. For those exponential growth of suburbs comes problems that seems to have been forgotten. Sea water intrusion for example; comes from a poor urban planning that allowed skyscrapers being built without environmental consent. The abundance of use of deep water also creates uncertain future for this city. Above all, the suburbs and the city itself was built incrementally without proper urban planning to create a sustainable city. The Economic interest seems to overrule the needs of creating sustainable and resilient environment. One thing for sure, the needs of mass transportation system is a must for Jakarta and the surrounding area. LRT is one option among other transportation model to overcome commuting problems surrounding Jakarta. The unique thing about LRT in Jakarta, it is not constructed by the government as part of their duty, instead it was proceeded through Public and

O. Hendra (✉) · L. S. Barus
Department of Urban Studies, School of Strategic and Global Studies, University of Indonesia, Depok City, Indonesia
e-mail: hendraoktia@gmail.com

E. M. Nia et al. (eds.), *Sustainable Development Approaches*, Lecture Notes in Civil Engineering 243, https://doi.org/10.1007/978-3-030-99979-7_19

Private Partnership. Here comes the next question of how does the conflict between economic interest and urban development arise and how do we construct the solution.

Keywords Light Rail Transit · Transit oriented development · Urban infrastructure development · Economic interest · Jakarta

1 Introduction

Light Rail Transit (LRT) has been developed by the Presidential Regulation No. 98, 2015 and endorse through Presidential Regulation No. 65, 2016. Under this regulation, the project was assigned to State Owned Company. Routes had also been assigned from Bekasi Barat to Cawang and From Bogor-Cibubur- Cawang then continue from Cawang to Kuningan Jakarta. The Routes itself was set in order to overcome traffic, derives from commuters from Bekasi and Bogor in access to Kuningan Jakarta every day. Since Jakarta is the capital city of Republic of Indonesia as well as centers for national economy, numbers of commuters from Jakarta, Bogor, Depok, Tangerang and Bekasi commutes in and out Jakarta every day. According to BPS [1] Jakarta branch in 2019, the number of commuters reached 3.2 million. Among this number only 27% are using public transportation. Therefore, having LRT will expectedly will decrease numbers of traffic form personal vehicles. Furthermore, LRT will be able to carry hundreds of thousands of people to reach their daily activities.

In order to support community's activities in the midst of capital density, the development of areas with mixed functions (M*ixed*-Use *Development)* can be an alternative for the Government in spatial arrangement. *Mixed-Use Development* is an area that is built by having several mixed functions such as occupancy, trade, office and commercial. The concept is supported by the Central Place theory explaining that the main purpose of a settlement and trade center is to provide for the needs of the surrounding population [2]. With the complexity, the hope is that connectivity can be realized, namely between the function of the building with the accessibility of the region. The concept of *Transit Oriented Development* is an area developed by combining *Mixed-Use* functions in order to make it easier for people to travel by foot, bicycle. Public transport [3]. This paper elaborates further whether the LRT Development from Jakarta- Bogor-Bekasi is being suspended because of the project is self or miss-conception of TOD surrounding the station that makes it difficult to reach by the commuter.

Light Rail Transit or commonly referred to as LRT is one type of rail- based public transportation that is included in the type of fast and light train. LRT is in high demand in major cities in the world whose function is as a *feeder* against larger-capacity trains, such as train (KRL) and MRT. In Indonesia, as the capital of the country that serves as the center of government as well as the center of the economy, Jakarta is in dire need of fast and cheap mass transportation. Commuters domiciled in cities around Jakarta, must be able to be integrated to support the mobility of their people. The presence of LRT as one of the choices of people's daily modes of transportation is highly anticipated, considering the rail-based mode of transportation that connects Jabodetabek only train (KRL).

Based on the Regulation of the Minister of ATR BPN No. 16 of 2017, the Principle of Transit Oriented Development or hereinafter referred to as TOD is the development of the region in order to encourage a sustainable-mobility through increasing utility of mass public transit and the development of environmental facilities of non-motorized and pedestrian, integrated with transit nodes.

Determining the location or point of the TOD Area is done through several stages, namely by determining the location of the potential area of TOD, the determination of the typology of the TOD area before deciding this location as the ultimate TOD area.

In supporting the choice of public transportation modes to reduce the traffic ratio in Jakarta, it is not sufficient by only providing adequate public transportation, but also requires a city development concept that provides convenience and comfort for road users, including zoning, passenger flow and intermodal integration. Recent studies attempted to find evidence of a relevance between TOD and transit location [4].

Determination of the location of potential area for TOD is done by making several studies such as; the study of the development of mass transportation systems, the study of the needs and direction of urban development, environmental studies, study of the carrying capacity of regional infrastructure, study of the characteristics of space utilization as well as the study of socio-economic conditions of the community. Spatially, areas that have the potential to become TOD, at least need to have the criteria to be in the transit node of the rail-based high-capacity mass public transit network. This area also meets the requirements of intermodal transit and in accordance with the direction of the development of service centers and activities. The determination of typology of the TOD is carried out based on the scale of mass transportation system services, the development of service centers and activities developed. Typology of the TOD area in question consists of the City TOD Area, Sub city TOD Area and Environmental TOD Area. The Determination of the Location of the TOD Area is stipulated in the Regional Regulation on Provincial/Regency/City RTRW.

Based on the TOD Standard conducted by ITDP [5], the TOD Principle in a region require to meet the following criteria:

1. Building an environment that supports running activities;
2. Giving priority to non-motorized transportation networks;
3. Creating a dense network of roads and pedestrian paths;
4. Placing development near high-quality mass transit networks;
5. Planning development with land use, income, and demographics mixed;
6. Optimizing space density and adjusting public transport capacity;
7. Build areas with a short travel distance;
8. Improve mobility through parking arrangements and road use policies.

2 Methods

This study uses a descriptive analysis approach. Researchers analyzed the existing conditions of LRT development and TOD area in Jabodetabek which impacted sustainability and economic aspects in the region. Data collection techniques are carried out by way of literature studies against the focus of research. After all data is obtained by the researcher, the research is conducted by adjusting existing conditions to the relevant regulations existed. Identify potential problems in the focus of studies, based on literature and spatial condition of the TOD. In order to explain financial interest, projection cash flow for both LRT and budgets for TOD is also being run in order to see whether the delays of project are related in certain ways.

3 Result Discussion

The appointment of Jabodetabek LRT development project is regulated in Presidential Regulation No. 98 of 2015 then the first amendment of Presidential Regulation No. 65 of 2016 and the second amendment of Presidential Regulation No. 49, 2017. PT. Adhi Karya (Persero) Tbk. under the regulation is assigned to build LRT infrastructure, namely lines (including overpass construction), stations, operating facilities and depots. PT. Kereta Api Indonesia (Persero) was appointed as management of this facilities, including procurement, operation, maintenance, business activities in surrounding area and the implementation of automatic ticketing systems.

Jabodetabek LRT already has a line that is contained under The Minister of Transportation Regulation No. 54 of 2013 on The General Plan of Mass Transit Network in Jabodetabek Urban Area. The first phase of LRT construction covers a total line length of 44 Km. The station is built as many as 17 stations with cross- service Cawang—Cibubur, Cawang—Dukuh Atas and Cawang—East Bekasi. The second phase covered the total length of the 39 Km line with 8 station points and 3 cross-service, namely Dukuh Atas—Senayan, Cibubur—Bogor, and Palmerah—Grogol (Fig. 1).

We run cash flows projection on Both LRT Project and a TOD in simple method. The calculation being made in order to see correlation between both projects to accelerate Return on Investment (ROI).

The presence of LRT is expected to be a solution for breaking-down congestions in Jabodetabek, since LRT is one of the rail-based modes that can replace the movement of private vehicle users to mass public transport. This is supported by affordable tickets for all people. In accordance with the world standard for mass transportation modes, the tickets are set around 1 USD. For these purposes we set the price for one LRT trip about IDR 15,000. LRT is expected to operate for 16 hours with capacity of 300,000 passengers per day. Under the assumption that:

- Daily Passenger Number: 300,000 Passengers.
- Amount of Daily Ticket Revenue: IDR 4,500,000,000.

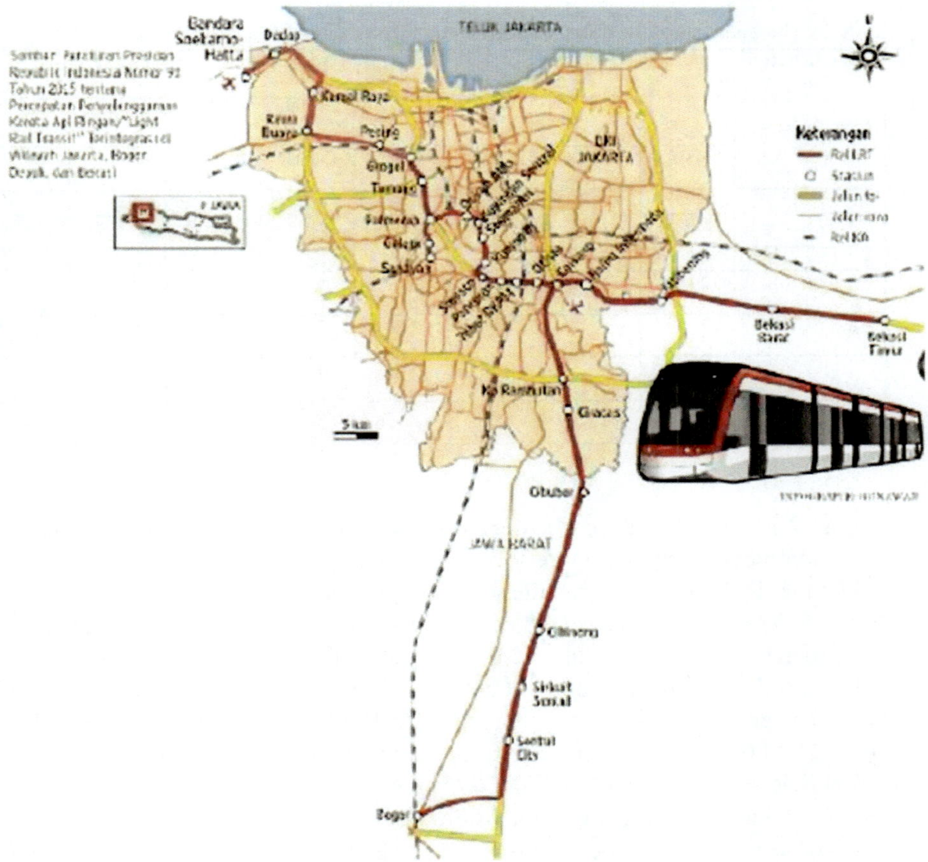

Fig. 1 Jabodetabek Transportation management agency *Source* Published Kompas.com 9 *September 2015*

- Amount of Ticket Revenue per Year: IDR 1.6 trillion.

Revenue earned by the operator, in addition to being sourced from daily ticket sales derives from advertising, merchandise and commercial rental areas. The investment cost incurred for the procurement of Jabodetabek LRT according to the Financial and Development Supervision Agency (BPKP) is worth 20 trillion Rupiah with the following details:

- Jabodetabek LRT Line: IDR10,480,000,000,000.00
- Station: IDR 3,710,000,000,000.00
- Operating Facilities: IDR 6,550,000,000,000.00

The Financial Projection Based on Investment Value and assumption above is described as follows (Table 1):

With total construction cost amounting IDR 25 trillion, Internal Rate of Return (IRR) reached 8,26%. It also means that the project will reach break-even point

Table 1 The financial projection investment value based on ticket's tariffs of year 2021

Construction Cost (Rp. Mil)	Tarif (Rp)							
		6.000	10.500	15.000	19.500	24.000	28500	33.000
	5,69%	40,00%	70,00%	100,00 %	130,00 %	160,00 %	190,00 %	220,00 %
3.593.317	10,00%	15,16%	23,46%	31,81%	40,41%	49,31%	58,50%	67,96%
14.373.267	40,00%	5,29%	9,62%	13,03%	16,03%	18,95%	21,86%	24,79%
25.153.217	70,00%	1,72%	5,58%	8,26%	10,53%	12,56%	14,42%	16,21%
35.933.168	100,00 %	-1,41%	3,27%	5,69%	7,64%	9,35%	10,90%	12,35%
46.713.118	130,00 %		1,56%	3,97%	5,76%	7,28%	8,66%	9,92%
57.493.068	160,00 %		0,01%	2,64%	4,38%	5,80%	7,05%	8,20%
68.273.018	190,00 %		-1,82%	1,50%	3,27%	4,64%	5,83%	6,89%

within 12.1 Year. Therefore, the development of the TOD area at each point of the LRT station becoming attractive solution for the investment.

For the TOD area, we run financial budgets and projection of a TOD area in Bogor. E Area stands above 1,2 hectares land. Assuming heights allowed (KKOP) of 70 meters, KDB of 0.3 and KLB 4, Developer will be able to build gross area of 112,667.3 m^2 TOD area and 106660.58 m^2 semi gross sale-able area. Assuming that price per m^2 for semi gross area reach Rp 18,000,000.00 and cost to build reach Rp. 9,900,000.00 per m^2 then the gross profit will be at Rp. 804.484 billion (1,919.890 billion IDR – 1,115.406 billion IDR). This gross profit will not include incomes from commercial area (mall, convenience store etc.).

Considering this income, building the same 10 Models of TOD along with LRT will be able to decrease investment about 8 trillion Rupiah, and this amount will be able to lower down cost of LRT investment in order to make it feasible for being developed under Public and Private Partnership scheme.

Since the development of the TOD area becomes important for developers that is assigned to Build LRT, Government should include all TOD Package in offering other project infrastructure. There are 2 important factors financially that has to be remembered by the developer:

1. There will be a huge economic of scale involving when building both TOD and LRT at the same time. A solid financial support and structures has to be prepared before taking this project.
2. Time management has to be determined in advance when executing this project combination. Miss conception in timing will create uncertainty in cost of run and it can harm the entire project.

Finally, it should be kept in mind that TOD is a means for support in reducing cost of development of LRT, not the main purpose. LRT is still the main goal to overcome commuting problems.

4 Conclusions

Infrastructure development in supporting any urban issues, as a matter of fact, can be solve by Private and Public Partnership. It is understood that it is the obligation of government to provide all of mass transportation system for its people. However, the government arm is limited to yearly budget that is established under the law. In order to expands projects that is purposed for development of infrastructures government can assigned part of its duty to Private Sectors. In the other hand, private sectors always seek for profit and economic interest under their operation. CRS that is made by the private sectors relatively small compare to economic value they normally gain.

In order to make the scheme work, the capacity and capability of private sectors that is assigned to build the infrastructure has to be assessed. This company has to be able to provide large amount of investment and build both main project and it supports at the same time. By this means, the infrastructure will be well built-in proper procedures and the profit targeted by the company can be achieved.

References

1. Statistic Central Bureau Jakarta Branch (2019) Commuter statistic for Jabodetabek 2019. CV. Ramsy Sinar Sejahtera
2. Christaller W (1933) Central places in Southern Germany. Germany
3. Calthorpe P (1993) The next American metropolis: Ecology, community and the American dreams. Pricento Architectural Press, New York
4. Purba R, Barus L (2022) The choice of transportation mode in transit location (Case Study: Dukuh Atas Area) University of Indonesia
5. Institute for Transportation and Development Policy (2017) TOD Standard, 3rd edn. ITDP, New York
6. Regulation of the Minister of Agrarian affairs and Spatial Planning/Head of the National Land Agency of the Republic of Indonesia Number 16 of 2017 on Guidelines for the Development of Transit-Oriented Areas. Jakarta: Agrarian and Spatial Planning/National Land Agency of the Republic of Indonesia